# The Making of Experimental Economics

Andrej Svorenčík • Harro Maas
Editors

# The Making of Experimental Economics

## Witness Seminar on the Emergence of a Field

Springer

*Editors*
Andrej Svorenčík
Department of Economics
University of Mannheim
Mannheim, Germany

Harro Maas
Centre Walras-Pareto d'études interdisciplinaires de la pensée économique et politique - IEPI
Université de Lausanne
Lausanne, Switzerland

ISBN 978-3-319-20951-7 ISBN 978-3-319-20952-4 (eBook)
DOI 10.1007/978-3-319-20952-4

Library of Congress Control Number: 2015947288

Springer Cham Heidelberg New York Dordrecht London


Printed on acid-free paper

Springer International Publishing AG Switzerland is part of Springer Science+Business Media (www.springer.com)

# Contents

# Chapter 1
# A Witness Seminar on the Emergence of Experimental Economics

**Harro Maas and Andrej Svorenčík**

> *Economists cannot make use of controlled experiments to settle their differences; they have to appeal to historical evidence, and evidence can always be read both ways.*
> *"What Are the Questions?"* (Joan Robinson, *1977)*

## Introduction

On May 28 and 29, 2010, eleven experimental economists gathered at the premises of the Royal Netherlands Academy of Arts and Sciences (KNAW) to participate in a so-called witness seminar on the history of the experiment in economics. The seminar was organized by Harro Maas and Andrej Svorenčík, principal investigator and Ph.D. student on a grant project that was funded by the Netherlands Organisation for Scientific Research (NWO) on the history of observational practices in economics.[1]

Some of the participants, like Vernon Smith and Charles Plott, will not have been surprised by our invitation. Others, like Jim Friedman or John Ledyard, perhaps more so as they did not and do not consider their main contribution to economics to be in experimental economics and may well have thought of someone else as a better candidate. Yet all except one agreed to participate in this event, moderated by the British experimental and behavioral economist Chris Starmer,

An erratum to this chapter can be found at DOI 10.1007/978-3-319-20952-4_10

[1] The witness seminar was made possible by generous funding of the Netherlands Organisation for Scientific Research (NWO), VIDI-research grant 276-53-004. Briefly, the grant project consisted of three subprojects that took sites of observing as its starting point: the observatory, the laboratory, and the armchair. On the observatory and the armchair as sites of observing, see D'Onofrio 2013 and forthcoming; on the armchair see Maas 2011. See also Maas and Morgan 2012.

H. Maas
Centre Walras-Pareto d'Etudes Interdisciplinaires de la pensée économique et politique, IEPI, Université de Lausanne, Bâtiment Géopolis, 1015 Lausanne, Switzerland
e-mail: Harro.Maas@unil.ch

A. Svorenčík (✉)
Department of Economics, University of Mannheim, L7 3-5, 68163 Mannheim, Germany
e-mail: svorencik@uni-mannheim.de

 
A. Svorenčík, H. Maas (eds.), *The Making of Experimental Economics*,
DOI 10.1007/978-3-319-20952-4_1

that took place over 2 intensive days in a stately room of the premises of the KNAW.

The seminar was audio and video recorded. These tapes, with all concomitant documentation, have now been deposited in the David M. Rubenstein Rare Book & Manuscript Library at Duke University as part of its ongoing *Economists' Papers Project* and are accessible for scholarly investigation. This book contains the fully edited and annotated transcript, short biographies of the participants, and a slightly revised background paper that Andreas Ortmann wrote in preparation of the seminar. Our introduction presents a discussion of the witness seminar method, an account of our preparations, what we think can be learned from it, and a consideration of its limitations.

## The Method of the Witness Seminar

A witness seminar is a moderated group conversation on a specific topic that was introduced as a method of historical inquiry almost simultaneously at the Wellcome History of Twentieth Century Medicine Group and the Institute for Contemporary British History in the early 1990s.[2] It has been tried at several other places afterward as well, for example, at the Royal Institute of Technology in Stockholm as a tool to uncover the history of IT in Sweden (Lundin, 2009).[3] One of the primary motivations for a witness seminar is to record memories that otherwise will be irrevocably lost.

There is no standardized way to conduct a witness seminar, at least not to date. Witness seminars can be devoted to strictly circumscribed events in place and time, such as the "Winter of Discontent in 1978-1979" (1987) or "Let Us Face The Future: the 1945 Labour General Election Victory" (5 July 1995), both run by the Institute for Contemporary British History, but also seminars on events spanning several decades, such as "The Bretton-Woods Exchange Rate System 1944-72" (1994).[4] In the history of science, technology, and medicine, they have been used as

[2] See Tansey's work for useful accounts of the method of the witness seminar. E. M. Tansey initiated and organized (and still does so) the witness seminars at the Wellcome History of Twentieth Century Medicine Group. **Tansey, E. M.** 2008. "The Witness Seminar Technique in Modern Medical History," H. Cook, A. Hardy and S. Bhattacharya, *History of the Social Determinants of Health.* Orient BlackSwan, 279–95, **Tansey, E. M.** 2006. "Witnessing the Witnesses: Pitfalls and Potentials of the Witness Seminar in Twentieth Century Medicine," R. E. Doel and T. Söderqvist, *The Historiography of Contemporary Science, Technology, and Medicine: Writing Recent Science.* London; New York: Routledge.

[3] To document the history of IT in Sweden, the science museum in Stockholm had organized two witness seminars that they combined with extensive face-to-face interviews of people involved in that history.

[4] For a complete list of witness seminars organized by the Institute for Contemporary British History at King's College, see https://www.kcl.ac.uk/sspp/departments/icbh/witness/WitSemscomplete.aspx [Accessed March 31, 2015]. The Winter of Discontent was a very cold

an entry for a better understanding of scientific discoveries and technological or institutional innovations that similarly may span several years if not decades or as a tool to uncover untapped archival resources. Examples of seminars run by the Wellcome History of the Twentieth Century Medicine Group are "Ashes to Ashes: The History of Smoking and Health" (1995), "Clinical Research in Britain, 1950-1980" (1998), or "Beyond the Asylum: Anti-psychiatry and Care in the Community" (2003).[5]

A witness seminar aims to bring together key participants of an important historical event to obtain a mix of different perspectives that may agree or disagree, but preferably lead to an exchange of memories that feed upon one another in interesting and unexpected ways. Participants not necessarily have to be high-profile individuals—Nobel laureates, politicians, and administratives high up the ladder—though in some cases it is difficult to conceive a seminar without them. The first witness seminar organized by the Wellcome History of Twentieth Century Medicine Group was about the discovery of monoclonal antibodies for which César Milstein, an Argentinean-born scientist, and Georges Köhler, his German postdoctoral fellow, received the Nobel Prize in Physiology and Medicine for 1984 (together with Niels Jerne). The seminar was especially organized to uncover the context around their conscious and controversial decision not to patent their invention (Tansey, 2006).

Clearly, the presence of these two key participants was crucial to the seminar and turned out to be even more so as both died within a few years after the seminar took place, thus preserving information that would have been permanently lost otherwise. But high-profile individuals tend to have been overexposed to the media and so maintain an almost scripted version of past events that make their contributions hardly go beyond what is already known from other (published) sources. The experience from both the Wellcome and Contemporary History seminars is that the communal setting may help to lift such individuals out of their standard grooves, though this of course not always happens (Tansey, 2006).

This is not to suggest, of course, that a witness seminar is intended to establish "the truth" of an event in any sense. Well known are the objections to oral history writing about biased and distorted memories, and such objections hold for the setting of the witness seminar as well and perhaps even more so. Because of the

---

winter in which Britain was haunted by a sustained period of strikes that eventually brought Margaret Thatcher to election victory. The name refers to Shakespeare's Richard III: "Now is the winter of our discontent/Made glorious summer by this son of York...." See also **Hay, Colin.** 2010. "Chronicles of a Death Foretold: The Winter of Discontent and Construction of the Crisis of British Keynesianism." *Parliamentary Affairs*, 63(3), 446–70.

[5] The first two were published as **Lock, Stephen; L. A. Reynolds and E. M. Tansey.** 1998. *Ashes to Ashes : The History of Smoking and Health*. Amsterdam; Atlanta, GA: Rodopi, **Reynolds, L. A. and E. M. Tansey.** 2000. "Clinical Research in Britain 1950–1980." *Wellcome Trust*, 7(7). The third is archived at the Wellcome Library: http://search.wellcomelibrary.org/iii/encore/record/C__Rb1946919__SGale,%20Robert%20Peter.__P0,3__Orightresult__X1;jsessionid=CFA74E92884D5DE50E02B3172212F614?lang=eng&suite=cobalt [Accessed on March 31, 2015].

communal setting, individuals may feel inhibited rather than encouraged to speak up on specific issues, in which cases recorded pauses and silences may become more informative than what is actually said. Alternatively, the very fact that participants engage in a conversation in which memories start feeding upon one another may lead to a communal group vision that is at tension with published or unpublished (archival) materials. As well argued by Hoddeson, rather than counting against oral histories or against this specific form of oral history, such tensions between oral and archival sources can provide fruitful nods that may lead to a reconsideration of interpretations and meanings that have been given to past events or that may lead to further investigations (Hoddeson, 2006). These may include substantial interpretative revisions of existing archival materials or the search for new archival sources.

This does not mean that commonly pronounced objections against oral history are invalid. But they not only hold against oral history, including the collective setting of a witness seminar. Archives are sometimes demonstrably incomplete or provide an image of events that hinges on the ordering of the archive and so run into the same kinds of difficulties that surround oral history sources. The reordering of the Alfred Marshall archives in Cambridge was also a reordering of the man and his work. The editions of the complete works and correspondence of David Ricardo, John Stuart Mill, or John Maynard Keynes created the persons as much as they conserved them. If an archive of a prominent individual enters special collections fully cataloged, that should be source of worry rather than joy for the historian. There is no reason to hold archival sources for the gold standard of historical evidence, and certainly for contemporary history, writing historians would do wise to cast their net a bit wider than the resources they traditionally have worked with.

This certainly holds for the history of economics where innovations in the documenting and writing of history so far have met with considerable skepticism. Ivan Moscati even went to so far to limit the historian of economics' field of vision to published sources only, thus straitjacketing the topics that can be studied and limiting the audiences that the history of economics might address (Moscati, 2008). This is especially important when it comes to the writing of contemporary history, for which important actors are still with us. Arjo Klamer's *Conversations with Economists: New Classical Economists and Opponents Speak Out on the Current Controversy in Macroeconomics* still stands out as a stellar example of an oral history endeavor that not only conserved but deliberately intervened in current debates (Klamer, 1984).

Within the history of economics especially Roy Weintraub has persistently advocated the use of oral resources to gain access to aspects of the development of the economics discipline that may remain hidden otherwise (Weintraub et al., 2007; Weintraub et al., 1998). Ross Emmett explains that one of the reasons he started his extensive oral history project on the Chicago Economics Department was to modify existing, sometimes self-gratifying stories of the success of the so-called Chicago school of economics (Emmett, 2007). Emmett interviewed individuals on a one-to-one basis. In addition to the great names, he interviewed students who did not finish their Ph.D. or minor economists, administratives, and other actors who might shed light on the role of funding agencies, institutionalized structures as the so-called Chicago workshop system, or on the interaction between the faculty and students or faculty and (local) government agencies and business.

While George Stigler may have held that the "details of a man's personal life" only serve to "distort ... the understanding of scientific work" (Emmett, 2007, p. 172), Emmett's oral history project confirms contemporary work in history and sociology of science that has convincingly shown that it is exactly such details, broadly taken, that help to understand the development of a discipline, including the individual's scientific work. The idea that the sources of scientific work are irrelevant to understanding is a figment of the mind, a relic of worn versions of the "view from nowhere" (Nagel, 1986). Historical or methodological studies that merely focus on the context of presentation—the published texts—rather than on the context of research practices simply fall short of contemporary standards in understanding the business of science, and economics is no exception. This pertains also to scientific developments of which the actors are still with us.[6]

## Background to the Witness Seminar on the Experiment in Economics

Our decision to organize a witness seminar on the experiment in economics was certainly motivated by our firm conviction that the history of a discipline does not map to its published sources, but there were other substantial reasons to organize the seminar as well.

The introduction of the experiment in economics produced, by any standards, a major change in the economist's research practice that goes beyond methods and methodology used. Joan Robinson was not alone in denying the experimental

---

[6] See especially **Doel, Ronald Edmund, and Thomas Söderqvist.** 2006. *The Historiography of Contemporary Science, Technology, and Medicine: Writing Recent Science*. London; New York: Routledge, **Söderqvist, Thomas.** 1997. *The Historiography of Contemporary Science and Technology*. Amsterdam: Harwood Academic.

For very perceptive essays on the problems and issues surrounding the writing of "the history of now," see **Hughes, Jeff.** 1997. "Whigs, Prigs and Politics: Problems in the Historiography of Contemporary Science." *The Historiography of Science and Technology*, 19–37; **Lewenstein, Bruce V.** 2006. "The History of Now: Reflections on Being a "Contemporary Archivist"; R. E. Doel and T. Söderqvist, *The Historiography of Contemporary Science, Technology, and Medicine: Writing Recent Science*. London and New York: Routledge; **Weintraub, E. Roy.** 2005. "Autobiographical Memory and the Historiography of Economics." *Journal of the History of Economic Thought*, 27(1), 1–11, ____, 2010; "Breit and Hirsch, Eds., Lives of the Laureates: Twenty-Three Nobel Economists." *History of political economy.*, 42(4), 779–82, **____.,** 2007. "Economists Talking with Economists, an Historian's Perspective," P. A. Samuelson and W. A. Barnett, *Inside the Economist's Mind: Conversations with Eminent Economists*. Malden, MA: Blackwell Pub., 1-11; **Weintraub, E. Roy and Evelyn L. Forget.** 2007. *Economists' Lives: Biography and Autobiography in the History of Economics*. Durham; London: Duke University Press.

method to economics. Way up into the 1980s, one could find in Paul Samuelson's and other textbooks statements in which it was emphasized that the specific nature of economic phenomena and the concomitant lack of a dedicated space of experimentation precluded the controlled experiment from the toolbox of the economist. In Samuelson and Nordhaus, we read:

> One possible way of figuring out economic laws ... is by controlled experiments. ... Economists [unfortunately] ... cannot perform the controlled experiments of chemists or biologists because they cannot easily control other important factors. Like astronomers or meteorologists, they generally must be content largely to observe. (Samuelson and Nordhaus, 1985, p. 8)

Apart from the fact that Samuelson and Nordhaus subscribe to a distinction between experimentation and observation that is at least problematic (Daston and Lunbeck, 2011; Maas and Morgan, 2012), they made their pronouncement at a time when experimental economists had been practicing their trade, with increasing intensity and in a rapidly growing community of experimentalists, for some 20 years or so.

Around the same time, that is halfway the 1980s, some of the more high-profile members of this growing community started reflecting on the novelty of their endeavors in papers, essays, and handbooks. They started considering the establishment of a separate community of like-minded economists, and they started discussions about the pros and cons of a separate journal for publishing experimental research. At the beginning of this new century, philosophers of (social) science turned toward a systematic investigation of the philosophical and methodological questions experimental economics posed.

Especially Francesco Guala's pathbreaking book on the methodology of experimental economics in 2005 opened up a new and still growing independent field of philosophical and methodological reflection on economic experimentation (Guala, 2005).[7] But Guala was well aware of the limitations of his book. Even though some chapters built on Guala's earlier published work in which he engaged with rich historical case material, he noted that at the time his book was published, no history of the emergence of the experiment in economics existed.

This was still the situation in early 2009, when we decided to organize our witness seminar. By then some good historical work had been published, for example, by Edward Nik-Khah and Kyu Sang Lee, but their work, understandably, zoomed in on particular case studies, such as the FCC auction that had also been studied, from a different angle, by Guala, or on particular individuals, such as the 2002 winner of the Sveriges Riksbank Prize in Memory of Alfred Nobel, Vernon Smith (Lee, 2004; Lee and Mirowski, 2008; Nik-Khah, 2006, 2008).

[7] See, for example, **Bardsley, Nick; Robin Cubitt; Graham Loomes; Peter Moffatt; Chris Starmer and Robert Sugden.** 2010. *Experimental Economics: Rethinking the Rules*. Princeton: Princeton University Press, **Santos, Ana Cordeiro dos.** 2009. *The Social Epistemology of Experimental Economics*. London: Routledge. Also issues of *Economics and Philosophy* and *Journal of Economic Methodology* are peppered with contributions on the methodological questions posed by the experiment in economics.

Lee was able to study Vernon Smith's work in some detail because Smith had donated his personal archives to the Economists' Papers Project at Duke University. Put somewhat uncharitably, Lee followed the historians' backfall option: go to the archives available, see what is there, and write a good story—and this is not to denounce such work; on the contrary, historians all do it and to a certain extent necessarily so. But for a better understanding of the emergence of the experiment, and the establishment of the laboratory as a dedicated site of inquiry, clearly more was needed than the scholarly work published until then. We had to create our archives rather than to rely on the scarce existing ones.

As we were particularly interested in the changes that the emergence of the experiment and the laboratory as a dedicated site of research had produced in the economists' research practices, a lack of sufficient materials was potentially damaging to this part of our grant project, and thus the potential that a witness seminar might offer in uncovering new or untapped resources was an important motive to push the idea forward. It was of course not clear if such materials, if uncovered, would prove merely helpful in writing a thick description of the emergence of the experiment in economics or if such materials would help pose different kinds of historical and methodological questions that had been asked so far. We hoped to be able to use the witness seminar as a lever for this purpose, and we think it served its purpose well, not only for what we will focus on here, its transcript, but also because it served as an important building block for Andrej Svorenčík's Ph.D. thesis *The Experimental Turn: A History of Experimental Economics* (Svorenčík, 2015).

## Preparation of the Seminar

Several decisions had to be made to move forward. Whom to invite and why? What topics to cover and why? How to practically prepare for and organize the seminar (and when)? And what to expect as a result? We will discuss each of these issues in turn, although, unsurprisingly, they cannot be wholly separated from another. We will devote a separate section to the seminar's results.

## Whom to Invite and Why?

Thesis work of our colleague Floris Heukelom, recently published with Cambridge University Press as *Behavioral Economics: A History*, convincingly showed that the historical trajectories of what are nowadays referred to as experimental and behavioral economics had been very different (Heukelom, 2014). Following Heukelom's work, we decided to importantly restrict the scope of the seminar and so to exclude (the history of) behavioral economics from our considerations. There was an additional reason for this important restriction. We were interested to uncover the history of a new method of research in economics, but we were not interested in battles from the trenches on theoretical stances, especially not in diverging theoretical stances on rational choice theory. We were afraid (rightly or wrongly) that the inclusion of behavioral economics could lead to discussions on the virtues and vices of rational choice theory, discussions that, as shown by Heukelom, could be separated from discussions on the historical trajectory of the experimental method. Limiting the time horizon roughly to the middle of the 1990s gave an additional rationale to our decision. Thus, to ensure the focus of the witness seminar would be on the method—the experiment in economics—rather than on theoretical considerations about rational choice, we limited our field of vision to experimental economics.

For any conversation on the emergence of the experiment in economics, we considered it inconceivable if the pioneers Vernon Smith, Reinhard Selten, and Charles Plott would not be present. Selten was also important because we wanted to include an international perspective, as even a cursory acquaintance with the history of the experiment in economics shows that there had been a flourishing experimental community of economists in Germany long before there was one in the United States. This community was instigated by Heinz Sauermann in the late 1950s, and Selten was an early and its most prominent member.

With these names fixed, we subsequently included interconnected individuals with different characteristics who could shed light on the different topics we considered that needed to be addressed (on which more below). We also had to address nontrivial considerations such as whether individuals could get on sufficiently well with one another to engage in an open conversation. We thus narrowed down our initial list of 25 names to the 12 individuals we decided to invite to the seminar.

As some of the participants are of high age, there was of course the risk of cancellations. This fortunately did not happen between the moment we sent out the invitations, in May 2009, and the seminar itself, in May 2010. Only Reinhard Tietz had to cancel because of brief ill-health a few days before the seminar. Unfortunately, that reduced the number of European participants to two, turning the seminar into a largely U.S. affair.

Another important decision concerned the choice of the moderator of the seminar. As neither of the organizers is a native speaker of English, we obviously had to look for a native speaker who ideally would have sufficient knowledge of experimental economics, without being him- or herself part of the historical trajectory under discussion. Also, it was a conditio sine qua non that the moderator would have sufficient historical sensibilities to focus the contributions of the participants on their historical relevance, without drifting off to an exchange of arguments that might better suit a philosophy or economics seminar. We found ourselves extremely happy that Chris Starmer agreed to play this role, possibly also because Robert Sugden had spoken favorably with Chris about our project, having participated in its opening workshop.[8]

[8] "Observation in Natural and Social Sciences, Historically Considered," opening workshop of NWO project Observation in economics, historically considered, March 12–13, 2009, Hortus Botanicus, Amsterdam.

Chris Starmer had taught economic history with one of the leading British historians of economics, the late Bob Coats, at Nottingham. Because our choice of timeline excluded the separate trajectory of the emergence of the experiment in Britain, Chris Starmer seemed the ideal person to take on this role. Also in this case, nontrivial considerations about his loud and clear voice and his way of addressing fellow participants in a seminar (with which one of us had very positive experiences) enhanced our decision to invite him. Chris agreed on the condition that a sufficient number on our selected list would agree to participate, somewhat assuming, as he told us afterward, that this would never happen. We started our invitations with Vernon Smith, Charles Plott, and Reinhard Selten, and once having their commitment, we proceeded with the rest. Within a few days, all but one invitee had accepted to participate.

## What Topics to Cover and Why?

Nowadays there is a large literature in the history and sociology of science on the experiment and the experimental method.[9] We used some of the more pertinent texts in this literature to identify the areas to be covered at the witness seminar: community building, funding, skills and techniques, and the laboratory as a research site. These topics and the timeline served as a loose grid to choose our cast of characters: participants should be spread over these areas, and there should be sensible relations between participants in terms of pioneers, first-generation Ph.D. students, early presidents of the *Economic Science Association*, first editor of the journal *Experimental Economics*, economists on NSF-committees, etc. We also structured the succession of sessions at the witness seminar along these topics. Below we list the economists we invited ordered by the year of birth (in brackets). Short biographical information can be found in Chap. 8 that indicates our various reasons for inviting them: pioneers, first generation of Ph.D. students, a spread in backgrounds in public choice, game theory, engineering, editors of *Experimental Economics*, members of NSF panels, or other administrative

[9] The literature is too vast to cover here. Classic references are **Collins, Harry M.** 1985. *Changing Order: Replication and Induction in Scientific Practice*. London; Beverly Hills: Sage Publications, **Galison, Peter.** 1987. *How Experiments End*. Chicago: University of Chicago Press, ____. 1997. *Image & Logic: A Material Culture of Microphysics*. Chicago: The University of Chicago Press, **Knorr Cetina, K.** 1999. *Epistemic Cultures: How the Sciences Make Knowledge*. Cambridge, Mass.: Harvard University Press, **Latour, Bruno and Steve Woolgar.** 1986. *Laboratory Life: The Construction of Scientific Facts*. Princeton, N.J.: Princeton University Press, **Shapin, Steven and Simon Schaffer.** 1985. *Leviathan and the Air-Pump: Hobbes, Boyle, and the Experimental Life*. Princeton, N.J.: Princeton University Press.

For a recent reflection on histories of the laboratory, see **Kohler, Robert E.** 2008. "Lab History: Reflections." *Isis*, 99(4), 761–68, and subsequent articles in that focus section of *Isis* for further references. Histories of the experiment and the laboratory have predominantly investigated the experiment/laboratory in the natural sciences, in medicine, and in psychology.

positions. Thus, this list, and our reason for inviting them, also reflects our own rudimentary understanding of the history of experimental economics at the time.

1. Vernon Smith (*1927)
2. Reinhard Tietz (*1928)
3. Reinhard Selten (*1930)
4. Jim Friedman (*1936)
5. Charlie Plott (*1938)
6. John Ledyard (*1940)
7. John Kagel (*1942)
8. Elizabeth Hoffman (*1946)
9. Frans van Winden (*1946)
10. Charlie Holt (*1948)
11. Stephen J. Rassenti (*1949)
12. Alvin Roth (*1951)
13. Chris Starmer (*1961) (moderator)

## How to Prepare for and Organize the Seminar

To prepare for the seminar itself, we followed several tracks. From our reading in the witness seminars held by the Wellcome History of Twentieth Century Medicine Group and by the Institute for Contemporary British History, we took the idea that it could be useful to increase the historical awareness of the participants by having them read one or two position papers. In the end, and after many fierce discussions among ourselves, we asked Andreas Ortmann to write a short paper in which he would discuss a few "museum pieces" to trigger the participants' memories and used a comparison between a physics and biology laboratory from the sociologist of science Karin Knorr Cetina's widely acclaimed *Epistemic Cultures: How the Sciences Make Knowledge* (Knorr Cetina, 1999). Ortmann's selection covers a wide range of historical routes of experimentation on economic issues and on purpose includes some not performed by economists. A slightly rewritten version of his paper is included as Chap. 9.

Highly consequential for the seminar itself proved the in-depth interviews that Andrej held with all participants separately over the course of the year following the invitation. Apart from participants of the seminar, Andrej interviewed other experimental economists who were close to our invitees in relevant ways (as coauthors, supervisors, students, etc.). By the time of the witness seminar, all participants had been interviewed separately at least once, with an average duration of the interview of around two hours. At present the number of interviews with experimental economists stands at 63, with over 100 hours of recordings in total.

These interviews enabled us to create "thick CVs" that showed the particular strengths of the individual participants, points of contact with the other participants, and unexpected moments and events that we could use to give instructions to Chris Starmer on issues to be covered and whom to give the first word on specific issues.

We sent Chris these CVs a fortnight in advance of the actual seminar for his personal preparation. This does not mean Chris simply followed a list of scripted questions, but it did mean there was an informal list of fallback options to which he could take recourse once a topic was, or seemed, prematurely exhausted. During the breaks at the witness seminar, we used this information to discuss with Chris Starmer the points of entry for the upcoming sessions.

The interviews also helped to stimulate the memories of the participants, and it helped to make them aware of the importance of the archival materials they had stored during their lifetime. Some of the participants granted Andrej access to their personal archives in advance of the seminar; others did so afterward. And as a consequence of the seminar, some participants agreed to hand over their archives (or parts of it) to the *Economists' Papers Project* at Duke University.

There were a great many practical issues that we won't discuss in detail here, even though none of them is trivial. One can think about the choice of hotel, choice of venue for the event itself, choice of recording technology, and seating arrangement—all of them are important also because of budgetary considerations. We choose a great location for the seminar's dinner—so great that we returned far too late back in the hotel. A half a year in advance of the seminar, we invited Chris Starmer to go through these matters on site and to discuss the best way to physically organize the seminar. We divided the seminar itself into six sessions to allow a smooth transition from one topic to another and to incorporate sufficient flexibility to expand on one of them if that seemed appropriate from the enfolding discussion.

We would like to single out one of these practical issues for discussion, however, and that is the choice to have or not to have an audience present at the seminar itself. Both the Wellcome Trust and the Modern History Research Centre seem to be more lenient in this than we considered appropriate. We wanted to ensure all participants would feel as little inhibited as possible to express whatever they wanted to express. We estimated an audience would be a hindrance, especially if this audience would consist of experimental colleagues of the experimental economics center in Amsterdam, CREED. Spouses who were present were neither allowed inside of the seminar room, but attended all other activities.[10] At the same time, we compromised by being present ourselves and also inviting Andreas Ortmann and our historian and philosopher of economics colleague Marcel Boumans to be present—Ortmann because he wrote the position paper and Boumans because we wanted to have feedback on the event from a colleague of the history of economics community. We did not allow either of them to take notes to ensure that the participants would feel their privacy was secured. Clearly, this is an issue on which one can take a different stance, and whichever is taken will be of influence on the unfolding of the seminar itself.

[10] The spouses included Marcia Friedman, Martha Ann Talman (wife of Charlie Holt), Harriette Kagel, Marianna Plott, and Candace Smith.

## The Seminar and Its Results

Although we sometimes talked about the witness seminar as an "experiment," even in our communication with the participants, whatever one considers an experiment, a witness seminar intentionally lacks the controlled intervention of an experiment. There were episodes that came out of Andrej's one-to-one interviews, of which confirmation during the seminar would have been useful, but it would have been almost a violation of our own intention to only create the conditions in which participants would feel free to express their memories, rather than repeat what we considered their memories should be. The choice of topics, and the advance preparation with Chris Starmer, limited, but did not control, what participants could and would contribute.

This doesn't mean that we, as organizers, did not intervene in the unfolding of the witness seminar itself. Chris Starmer remembers our emphasis on the use of *surprise* in asking questions, such as asking about examples of *failed* experiments and telling something about them (in the session about "skills and techniques"). Surprise also was an important consideration in how we envisioned the start of the seminar. We considered it imperative that the first question would not be posed to Vernon Smith. And this was not just for the obvious reason that he might give a scripted answer, but that it could possibly confirm other participants in expectations about the seminar. We instructed Chris to ask the first question to Jim Friedman, which would most likely come as a surprise to the participants and thus, we hoped, would unsettle their expectations. We informed Jim Friedman about this briefly before the first session, and we also informed him about the substance of the question, which dealt with his decision to use experiments in his thesis. Chris would then move with a similar question to Reinhard Selten and only then to Vernon Smith. Afterward the floor was open to all participants. A few further remarks on these results are in order.

First, and as already indicated, the seminar helped to tap into a wealth of private resources, some of which are now in the (semi-)public domain, in the *Economists' Papers Project* at Duke University in particular. Among the papers deposited there due to the witness seminar (and Andrej's persistence) are the papers of Alvin Roth and Jim Friedman. Archives of several other experimentalists are still being negotiated. There are personal collections of papers Andrej got access to that will be deposited elsewhere, because they had been promised already as in the case of Elizabeth Hoffman's papers. Second, not in terms of archival materials, the witness seminar pointed us, for example, to unexpected actors such as Austin C. Hoggatt whose laboratory at Berkeley turned out to be an important reference point for Reinhard Selten and Vernon Smith and important for Jim Friedman's experimental work. Third, the witness seminar made the community of experimental economists aware of its own history, and this has led to some activity at international conferences.[11]

More important is whether the witness seminar changed and improved our understanding of the history and meaning experiments in economics. In the context of our grant project, we were particularly interested in issues that pertained to what counts as an observation in an economics laboratory setting. We do not intend to fob off the reader with a bromide, if we say the seminar directed us to material practices of control not, or not in sufficient detail, addressed in earlier scholarship. These can be almost trivial, such as making sure experimental subjects should "watch the screen" or that they should be enforced to read their payoffs in the intended direction, as reported by Plott. The seminar also made us realize how intricately interwoven epistemic and social cultures of experimentation are and how important it is to pay attention to the dates of events and episodes.

Smith, Ledyard, Kagel, and Plott are all connected via Purdue, but when Kagel came there at the end of the 1960s, Smith had already left and there was no institutional memory of the experiments Smith had performed in the early 1960s. Even during John Ledyard's studies at Purdue in the mid-1960s, he did not encounter experiments at all. Thus, Kagel and Smith's pathways into experimental economics were very different. Friedman, Smith, and Selten are all three connected to Hoggatt, but whereas Friedman's experiences with Hoggatt made him, more or less, give up on experiments, Smith and Selten were inspired by Hoggatt's laboratory and adjusted his example to their own needs once they had the opportunity to build a lab for their own. In addition, their experiences highlight that a state-of-the-art technological infrastructure is not enough, and not even a precondition, for the emergence of an experimental culture. Game theory was important for some, but not for all participants to become infected with the "experimental bug," as Plott put it.

Both Purdue and Caltech were important institutions for the emergence of the experiment in economics, which can also be seen as part of our selection bias of participants. However, the fact that Purdue and Caltech were not like traditional economics departments divided in separate groups (macro, micro, labor, etc.) appeared an important enabling condition for experimenting with experimentation. Carnegie Mellon, commonly less associated with experimental economics, offered Friedman a job,[12] Holt studied there, and Ledyard went there after graduating from Purdue. Just as at Purdue and Caltech, there was an openness to new ways of doing research not found at established economics departments, thus confirming studies in history of science which show that scientific innovation in great many cases comes from the (relative) fringes of a discipline.

---

[11] There has been a joint session at the ASSA between the History of Economics Society (HES) and the Economic Science Association (ESA) on the twenty-fifth anniversary of ESA (2012), a similar session organized by the ESA itself (2012), and another session on experimental economics' history at ASSA (2013) and the relationship of economics and psychology at ASSA (2015) that were also attended by experimental economists.

[12] Friedman's Yale classmate F. Trenery Dolbear who also wrote an experimental thesis was offered a job at Carnegie as well, but unlike Friedman he took it.

This should not lead one to conclude the economic mainstream was explicitly hostile to the experiment. On the contrary, the early experiences of Jim Friedman or Alvin Roth in getting their first experimental paper published in journals such as *Econometrica* and *Management Science*, or in receiving research funding (as in Friedman's case), gave them the illusion that their academic lives would be easy. This seemingly contrasts with the shared sense among the first generation of Ph.Ds. trained in experimental economics not to advertise themselves as such in the job market. Seemingly, because at the time there was no sustained separate teaching in experimental economics, and job interviews followed JEL code classifications in which experiments until the mid-1980s did not figure at all. At conferences, experimental papers were placed in educational sessions, which did not contribute to their standing. Publication seemed to become more difficult once the "experimental bug" had spread more widely in the economics community and referee reports began to become much more sophisticated than the "this is an experiment, that's interesting, let's publish it" experience of Roth. Such an experience almost cries for further research. From Svorenčík's subsequent work, acceptance and rejection policies in mainstream journals turn out to be a far more complex story, but it is exactly such further research that, we hope, will be prompted by this witness seminar (Svorenčík, 2015, Chap. 5 in particular). Again, institutional constraints as trivial as the JEL code may inhibit new techniques and methods to catch on. Institutional support, from NSF administrators such as Dan Newlon, market-sensitive editors of major presses, or contract research for companies who searched solutions for concrete problems, provided the necessary funding for subject payments as well as personal and technological infrastructure for the method to survive, stabilize, and grow.

A remark is in order on the relation between theory and data, and the role of the experiment in this relation, which led to some of the most heated discussions at the seminar. If we look at the characteristics of our participants, we see a strong predilection for theory. Most of the participants were theorists by training, more specifically mathematical economists, which was not typical in the 1950s and 1960s when many of them graduated. This may be seen to confirm received ideas that early experiments in economics were predominantly theory driven; intended to test theory. Attitudes to theory testing and how such a test should look like have differed however widely among participants. In addition, the relation between theory and experiment is much less straightforward or one directional than commonly assumed. Selten would not have gotten to the idea of sub-game perfect equilibrium, if it had not been for the experiments he had performed. Friedman left experimentation and turned into a theorist because of a lack of connection between his experiments and the theory he intended to test.

There were other themes explored during the seminar as well, for example, the evolution of the laboratory from the classroom to the virtual space of some of Plott's recent experiments or to the portable laboratories that can transform any laptop to a site for experimentation. Participants easily traced back the economists' emphasis on real incentives, not only to the well-known Friedman-Wallis critique but also to the psychologist Sidney Siegel, which may come as a surprise to most economists. Such themes indicate that the witness seminar not only opened up research venues that are predominantly of historical interest but also venues that bear on the philosophical and methodological reflections on the meaning and role of the experiment in economics. Thus, the witness seminar invites us to rethink the epistemic and sociological status of the experiment in economics.

## The Transcript

The transcript is preceded with a brief manual on how to read the transcript, the footnotes, and the endnotes. It is important to stress the difference between a spoken and a published text. A literal transcript would be, literally, unreadable. A rough first transcript was made by a professional company after we sampled several agencies to make an example of a transcript, which differed substantially the one from the other. Once we had a rough transcript, Andrej relistened the tapes and watched the videos and revised the text. Then he further edited the transcript by adding extensive footnotes and questions for participants where he deemed necessary, in loose consultation with Harro. This version of the transcript was then sent to the participants for approval and comments. Approval and comments do not mean control. In advance of the seminar, we asked all participants to sign an agreement that the original audio- and videotape, and a transcript of the event, would be made available for scholarly purposes. Approval thus only meant the participants agreed with the transcription, even though they would have phrased things differently on second thought. During the revision process, participants wanted to expand on their points or react to what others said. Such remarks and clarifying additions that were made after the seminar were on our discretion relegated to the endnotes, as explained in the reading guide of the transcript. The participants subsequently granted us permission for the present publication.

In literary studies, the question "what is a text" was and still is an important question, and it is pertinent to the present text as well. We deleted time markers and added subject headings, increasingly moving away from the experience of listening to the original tapes. The text that is published here thus also serves as an invitation to rethink the way we make our historiographical choices. In his introductory words, Chris Starmer mentioned the four partly overlapping themes that were to be discussed during the witness seminar, but only loosely, as he said, because "in many ways, it is for us to create the story for those themes as we go along" (this volume, p. 20).

Starmer's remark is very true. The text presented will not be the last say on the issues that were on the table, community building, techniques and skills, funding, and the laboratory as a site of research, but rather serves as a start. The text is a collectively created document in which first-person participants jointly reflected on a major event in the economics discipline, the introduction of the experimental method. But it tells only one story, and as Joan Robinson reminds us, history can be read in different ways. Thus, the transcript may serve as an entry to a fuller understanding of this episode in the history of economics and to the understanding of the experiment as a method of social inquiry. Hopefully, this volume will be an incentive to create other stories that need to be told.

## How to Read the Transcript of the Witness Seminar

The recording of the two days of the witness seminar lasts seven hours and fifteen minutes. This includes small parts made during breaks, which were not transcribed. The recording was originally transcribed in a verbatim fashion with some grammar editing. Then the recording was listened to again and the transcript was pruned of hundreds of filler words such as—ehm, uhm, well, I think, and, you know, kind of, I guess, I mean, of course, okay, etc. Whenever such words were used as qualifiers reflecting a state of uncertainty, confirmation of what others said, or groping in one's memory, they were retained. Further grammar errors were remedied and some stylistic changes to transform spoken word to written were introduced (e.g., breaking up long sentences, word order, nested ideas rearranged, etc.) while maintaining individual speaker's tone and authenticity of the transcript. Some additions to help the reader were made—missing subjects, unfinished thoughts—all inserted in square brackets [].

**Sounds**—most sounds were omitted in the transcript (coughs, tearing of paper, etc.)—all very infrequent. Hitting the desk, snapping fingers, and particularly laughs were retained—and recorded in several forms according to intensity—*laughter*, *some laughter*, and *group laughter*. There were only a handful of pauses made during the seminar. All are noted and inserted in square brackets [].

**Speakers**—are introduced by their full names. Two Johns and Charlies were present, and to avoid confusion, their surnames are inserted in square brackets whenever they are mentioned by name.

**Paragraphs**—long monologues are broken into paragraphs, which only infrequently reflect pauses made by the speakers.

**Cross talk**—In cases when someone made a short insertion, this is indicated in the text of the other person—in square brackets [].

**Names**—many people were mentioned during the seminar. Most have a short biographical entry on the following site: http://www.springer.com/de/book/9783319209517. Names that appear in the biographical appendix are in italics.

**Footnotes**—are used for three purposes. (1) They contain bibliographic references for papers referred to in the seminar. (2) They furnish details on alluded historical events and substantial issues. (3) They are answers to editor's follow-up questions. Continuous Arabic numbering is used.

**Endnotes**—include comments of participants made on the transcript. Continuous Roman numbering is used.

## References

Bardsley, Nick; Robin Cubitt; Graham Loomes; Peter Moffatt; Chris Starmer and Robert Sugden. 2010. *Experimental Economics: Rethinking the Rules*. Princeton: Princeton University Press.

Collins, Harry M. 1985. *Changing Order: Replication and Induction in Scientific Practice*. London; Beverly Hills: Sage Publications.

Daston, Lorraine and Elizabeth Lunbeck. 2011. *Histories of Scientific Observation*. Chicago: University of Chicago Press.

Doel, Ronald Edmund and Thomas Söderqvist. 2006. *The Historiography of Contemporary Science, Technology, and Medicine: Writing Recent Science*. London; New York: Routledge.

Emmett, Ross B. 2007. "Oral History and the Historical Reconstruction of Chicago Economics." *History of Political Economy*, 39, 172–94.

Engvall Siegel, Alberta. 1964. "Sidney Siegel: A Memoir," S. Messick and A. H. Brayfield, *Decision and Choice. Contributions of Sidney Siegel*. New York McGraw-Hill Book Co, 1–23.

Galison, Peter. 1987. *How Experiments End*. Chicago: University of Chicago Press.

____. 1997. *Image & Logic: A Material Culture of Microphysics*. Chicago: The University of Chicago Press.

Guala, Francesco. 2005. *The Methodology of Experimental Economics*. Cambridge; New York: Cambridge University Press.

Hay, Colin. 2010. "Chronicles of a Death Foretold: The Winter of Discontent and Construction of the Crisis of British Keynesianism." *Parliamentary Affairs*, 63(3), 446–70.

Heukelom, Floris. 2014. *Behavioral Economics: A History*. Cambridge University Press.

Hoddeson, L. 2006. "The Conflict of Memories and Documents: Dilemmas and Pragmatics of Oral History," R. E. Doel and T. Söderqvist, *The Historiography of Contemporary Science, Technology, and Medicine: Writing Recent Science*.

Hughes, Jeff. 1997. "Whigs, Prigs and Politics: Problems in the Historiography of Contemporary Science." *The Historiography of Science and Technology*, 19–37.

Klamer, Arjo. 1984. *The New Classical Macroeconomics: Conversations with the New Classical Economists and Their Opponents*. Brighton: Wheatsheaf.

Knorr Cetina, K. 1999. *Epistemic Cultures: How the Sciences Make Knowledge*. Cambridge, Mass.: Harvard University Press.

Kohler, Robert E. 2008. "Lab History: Reflections." *Isis*, 99(4), 761–68.

Latour, Bruno and Steve Woolgar. 1986. *Laboratory Life: The Construction of Scientific Facts*. Princeton, N.J.: Princeton University Press.

Lee, Kyu Sang. 2004. "Rationality, Minds, and Machines in the Laboratory a Thematic History of Vernon Smith's Experimental Economics," University of Notre Dame, 301.

Lee, Kyu Sang and Philip Mirowski. 2008. "The Energy Behind Vernon Smith's Experimental Economics." *Cambridge Journal of Economics*, 32(2), 257–71.

Lewenstein, Bruce V. 2006. "The History of Now: Reflections on Being a "Contemporary Archivist"," R. E. Doel and T. Söderqvist, *The Historiography of Contemporary Science, Technology, and Medicine: Writing Recent Science*. London; New York: Routledge.

Lock, Stephen; L. A. Reynolds and E. M. Tansey. 1998. *Ashes to Ashes: The History of Smoking and Health*. Amsterdam; Atlanta, GA: Rodopi.

Lundin, Per. 2009. "From Computing Machines to IT: Collecting, Documenting, and Preserving Source Material on Swedish IT-History." *History of Nordic Computing*, 2, 21–3.
Maas, Harro and Mary S. Morgan. 2012. *Observing the Economy: Historical Perspectives*. Durham; London: Duke University Press.
Moscati, Ivan. 2008. "More Economics, Please: We're Historians of Economics." *Journal of the History of Economic Thought*, 30, 85–91.
Nagel, Thomas. 1986. *The View from Nowhere*. New York: Oxford University Press.
Nik-Khah, Edward. 2008. "A Tale of Two Auctions." *Journal of Institutional Economics*, 4(1), 73–97.
____. 2006. "What the FCC Auctions Can Tell Us About the Performativity Thesis." *Economic Sociology–European Electronic Newsletter*, 7(2), 15–21.
Reynolds, L. A. and E. M. Tansey. 2000. "Clinical Research in Britain 1950–1980." *Wellcome Trust*, 7(7).
Robinson, Joan. 1977. "What Are the Questions?" *Journal of Economic Literature*, 15(4), 1318–39.
Samuelson, Paul A. and William D. Nordhaus. 1985. *Economics*. New York: McGraw-Hill.
Santos, Ana Cordeiro dos. 2009. *The Social Epistemology of Experimental Economics*. London: Routledge.
Shapin, Steven and Simon Schaffer. 1985. *Leviathan and the Air-Pump: Hobbes, Boyle, and the Experimental Life*. Princeton, N.J.: Princeton University Press.
Söderqvist, Thomas. 1997. *The Historiography of Contemporary Science and Technology*. Amsterdam: Harwood Academic.
Svorenčík, Andrej 2015. *The Experimental Turn: A History of Experimental Economics*. Ph.D. dissertation. University of Utrecht.
Tansey, E. M. 2008. "The Witness Seminar Technique in Modern Medical History," H. Cook, A. Hardy and S. Bhattacharya, *History of the Social Determinants of Health*. Orient BlackSwan, 279–95.
Tansey, E. M. 2006. "Witnessing the Witnesses: Pitfalls and Potentials of the Witness Seminar in Twentieth Century Medicine," R. E. Doel and T. Söderqvist, *The Historiography of Contemporary Science, Technology, and Medicine: Writing Recent Science*. London; New York: Routledge.
Weintraub, E. Roy. 2005. "Autobiographical Memory and the Historiography of Economics." *Journal of the History of Economic Thought*, 27(1), 1–11.
____. 2007. "Economists Talking with Economists, an Historian's Perspective," P. A. Samuelson and W. A. Barnett, *Inside the Economist's Mind: Conversations with Eminent Economists*. Malden, MA: Blackwell Pub., 1-11.
____. 2010. "Breit and Hirsch, Eds., Lives of the Laureates: Twenty-Three Nobel Economists." *History of Political Economy*, 42(4), 779–82.
Weintraub, E. Roy and Evelyn L. Forget. 2007. *Economists' Lives: Biography and Autobiography in the History of Economics*. Durham; London: Duke University Press.
Weintraub, E. Roy; Stephen J. Meardon; Ted Gayer and H. Spencer Banzhaf. 1998. "Archiving the History of Economics." *Journal of Economic Literature*, 36(3).

# Chapter 2
# The Very Beginnings

## Choosing a Dissertation Topic

Chris Starmer: Good afternoon and welcome. For the record, let me say today is May 28, 2010. My name is *Chris Starmer*, and I am the moderator of this Witness Seminar on the *Emergence and Evolution of Experimental Economics*. The event is organized by *Harro Maas* and *Andrej Svorenčík*, and funded by the *Dutch Science Foundation*. We are at the premises of the *Royal Dutch Academy*,[1] and together with me are, from my right, participants *Frans van Winden, John Ledyard, Jim Friedman, Charlie Holt, Vernon Smith, John Kagel, Betsy Hoffman, Reinhard Selten, Charlie Plott, Al Roth, and Stephen Rassenti.*[2] Welcome to all. During the event, my plan is, over a number of sessions, to explore with you four broad topics.

These are loosely defined topics relating to the community of experimental economists, issues relating to funding, skills and techniques of experimental economics, and issues related to laboratories of experimental economics. I am not going to

[1] The official name of the Dutch Science Foundation is The Netherlands Organisation for Scientific Research, the full title of the KNAW is *The Royal Netherlands Academy of Arts and Sciences*, or Koninklijke Nederlandse Akademie van Wetenschappen. It is housed at the Trippenhuis, Kloveniersburgwal 29, in Amsterdam. The seminar took place in *The Old Meeting Room* located on the first floor of the Trippenhuis Building with views out on the canal. The adjacent *Rembrandt Room* (so named because Rembrandt's famous Nightwatch covered one of its walls) was used for breaks.

[2] A twelfth participant Reinhard Tietz, a German experimental economist, was unable to attend the Witness Seminar due to an illness.



A. Svorenčík, H. Maas (eds.), *The Making of Experimental Economics*,
DOI 10.1007/978-3-319-20952-4_2

dwell at this point upon those themes, and I think in many ways, it is for us to create the story for those themes as we go along. The first session deals with issues related to the growth of the community of experimental economists, and I plan to start by asking two or three questions initially directed at particular individuals just to break the ice, so to speak, to the extent that the ice needs breaking. But then to very quickly move to open questions to which everybody is invited to contribute.

I am going to put the first question to Jim Friedman. It is related to your dissertation from Yale, in 1963, on the theory of oligopoly.[3] To my knowledge, it constitutes the first thesis within experimental economics.[4] I wonder if you can tell us something about the circumstances that led to the work in that thesis.

Jim Friedman Yes. My first exposure to anything experimental in economics was in the Fall semester of 1959 when I was taking a required microeconomic theory course from *William Fellner* as a beginning graduate student. In that course he had a couple of experimental papers. I think one of them was Vernon's JPE competitive markets paper,[5] and the other was *Mosteller* and *Nogee* who were not economists but did a utility experiment.[6] Following that, in my second year, I was taking a course during the whole academic year from a visiting professor who was *Martin Shubik*.

On toward the spring of that year, when I was floundering about coming up with a dissertation topic, *Martin* said, "Why don't you do an experiment?" He had talked a little about experimentation during the course that I had had with him. Following that, in September of '61, which was the beginning of my third year—that was the point when qualifying and comprehensive examinations were being done. In the oral part of those examinations, I was asked by somebody on the committee, probably *Koopmans*, what were my dissertation plans.

---

[3] Friedman's dissertation was titled *The Theory of Oligopoly*. An abridged version was published in: **Friedman, James W.** 1963. "Individual Behavior in Oligopolistic Markets: An Experimental Study." *Yale Economic Essays*, 3(2), 359–417.

[4] In the same year F. Trenery Dolbear, now at Brandeis University, also graduated from Yale. An abridged version was published in: **Dolbear, F. Trenery.** Ibid. "Individual Choice under Uncertainty: An Experimental Study." 419–69.

[5] **Smith, Vernon L.** 1962c. "An Experimental Study of Competitive Market Behavior." *The Journal of Political Economy*, 70(2), 111–37. And a correction ____. 1962a. "Errata: An Experimental Study of Competitive Market Behavior." *The Journal of Political Economy*, 70(3), 322–23.

[6] **Mosteller, Frederick and Philip Nogee.** 1951. "An Experimental Measurement of Utility." Ibid.59(5), 371–404.

I said, "Oh, I'm going to do an experiment," I think, but I had no concrete plans of any sort. *Tjalling* [*Koopmans*] says have I seen the work of *Siegel* and *Fouraker*. This was referring to their second book, the one called *Bargaining Behavior*,[7] which at that point was unpublished, but it was circulating as two fat working papers.[8] Of course, I had not seen it. I had not heard of them. He lent me copies, and it was the basis of what I did. That work taught me a lot about how to run an experiment, and it provided also a starting point from which I designed what I did.

Chris Starmer: As you initially described it, it seemed like the only option you had come up with. But you sound like you got more excited in the idea as you understood the work that was emerging.

Jim Friedman: Well, what I would say, Chris, is this. I had previously been thinking about various areas [of economics] in which I never really got my mind around something I could figure out how to do and that would seem to have some merit. Somehow, just instinctively, when *Martin* made the experimentation suggestion, I had this feeling—yes, I think I can do that. I can do something interesting. And I had that sense without having a precise plan about what to do. But the plan itself then followed on very readily once I absorbed the Siegel-Fouraker manuscript.

Chris Starmer: When did you become aware of Reinhard Selten's work on oligopolies?

Jim Friedman: [pause] I'm not sure. I know when I met Reinhard; we were just talking about that yesterday.[9] [It was] the summer or late spring of 1967. Summer to early fall was the end of a year that I had spent at Berkeley in which I had spent a great deal of time with *John Harsanyi*. It was [also] the beginning of a year when Reinhard was going to go to Berkeley and do the same thing. I spent a lot of time with *John*. But I would say he spent

[7] **Fouraker, Lawrence E. and Sidney Siegel.** 1963. *Bargaining Behavior*. New York: McGraw-Hill.

[8] **Siegel, Sidney; Lawrence E. Fouraker and Donald I. Harnett.** 1961. *Bargaining Behavior. Volume 1. The Uses of Information and Threat by Bilateral Monopolists of Unequal Strength.* Pennsylvania State University. **Fouraker, Lawrence E. and Sidney Siegel.** 1961. *Bargaining Behavior. Volume 2. Experiments in Oligopoly.* Pennsylvania State University.

[9] Both Friedman and Selten attended a dinner the night before the seminar. The other attendees were Marcia Friedman, Charlie Holt with wife Martha Ann Talman, Betsy Hoffman, John and Harriette Kagel, John Ledyard, Harro Maas, Andreas Ortmann, Charlie and Marianna Plott, Stephen Rassenti, Vernon and Candace Smith, and Andrej Svorenčík.

time there more effectively than I did.[10] [some laughter] But in any case, Reinhard stopped in New Haven [where Yale University is located.] I knew at least of him from *John [Harsanyi]'s* mention.[11] I don't know if I knew of him through other channels prior to that time. And then we met in the summer or early fall of '67.

Chris Starmer: Okay. Well, thank you. Perhaps I could turn to Reinhard because you published in Germany in 1959 the first paper on experimental economics by anyone sitting around this table. And I wondered how did you come up with the idea of doing an experiment?

## Crossing Disciplinary Boundaries

Reinhard Selten: [pause] When I began to study mathematics [at the Johann-Wolfgang-Goethe-University in Frankfurt am Main], I also went to psychology courses, and I actually studied a lot of psychology. The psychology students had to take part as subjects in experiments, at least at that time in Frankfurt, they had to have a certain number of points together in order to get later into the proseminar.[12] I gathered enough points and went also to the proseminar. I found the psychological experiments quite interesting. Afterwards, it was also always explained to us what the experiments were about. And so I got familiar with the technique of running experiments with human subjects. Not that I ran any on my own. I was just a subject. But that made me familiar [with them].

Later, I was doing game theory. After I finished my masters [in mathematics] on game theory,[13] I was hired by Professor *Sauermann* for a research project, which was about

---

[10] **Editor**: Is this a reference to Selten and Harsayni's work that led to their joint Nobel Prize? **Jim Friedman**: "No, I only meant that Reinhard and John clicked professionally in a way that John and I did not. John and I talked regularly, but we didn't get into any collaborative work and I don't think either of us had any significant impact on the other."

[11] Selten spent the academic year 1967–68 as a Visiting Full Professor at School of Business Administration, University of California at Berkeley. He met Harsanyi in 1961 when visiting Oskar Morgenstern, a good friend of his advisor Heinz Sauermann, in Princeton.

[12] A course of study for graduate and advanced undergraduate students, conducted in a manner of a seminar.

[13] Selten wrote his master's thesis in 1957.

application of decision theory to the theory of the firm. At this time, I had several things. First, I read *Herbert Simon* and got convinced of his ideas about bounded rationality.[14] Second, I saw a paper by—I don't remember the name of the author—the American Management Association edited a book about a computer management game.[15] I think it was the first, which was in the literature.

Also, I have read the seminal paper by Kalish, Millnor, Nash and Nehring,[16] the experimental paper from 1954 in the book edited by Thrall, Coombs and Davis *Decision Processes*. Of course, in my [doctoral] thesis, I have worked on cooperative games.[17] And I was quite impressed by the fact that it was possible to approach the questions of cooperative game theory by experiments.

Then I was very much interested in oligopoly theory. I read a lot about oligopoly theory. [After I had become aware of the computerized management game,] I thought that we can run oligopoly experiments and we do not need a computer for this purpose. What they can do with this computer, I also can do without it. I designed an oligopoly experiment with Cournot oligopoly [model] basically. After I had run this, my Professor *Sauermann* asked me to write it down. And I said, "Well, I don't know how to write an economics article [laughs], I have no feel for it." So we wrote it together.[18]

Chris Starmer Did you feel like you were doing something very unusual at the time?

[14] **Editor**: Do you remember which papers did you read? **Reinhard Selten:** "I have read the seminal papers:" **Simon, Herbert A.** 1955. "A Behavioral Model of Rational Choice." *The Quarterly Journal of Economics*, 69(1), 99–118. and ____. 1956. "Rational Choice and the Structure of the Environment." *Psychological review*, 63(2), 129–38., both reprinted in ____. 1957. *Models of Man: Social and Rational; Mathematical Essays on Rational Human Behavior in Society Setting*. New York: Wiley.

[15] **Ricciardi, Franc M.** 1957. *Top Management Decision Simulation: The A.M.A. Approach*. New York: American management association.

[16] **Kalisch, Gerhard K.; J. W. Millnor; John F. Nash and E. D. Nering.** 1954. "Some Experimental N-Person Games," R. M. Thrall, C. H. Coombs and R. L. Davis, *Decision Processes*. New York: Wiley, 513–18.

[17] Selten's dissertation: **Selten, Reinhard.** 1961. "Bewertung Von N-Personenspielen," Frankfurt: Johann Wolfgang Goethe-Universität Frankfurt am Main.

[18] **Selten, Reinhard and Heinz Sauermann.** 1959. "Ein Oligopolexperiment." *Zeitschrift für die gesamte Staatswissenschaft*, 115, 427–71. Reprinted in **Sauermann, Heinz.** 1967. *Beiträge zur Experimentellen Wirtschaftsforschung*. Tübingen: Mohr., pp. 9–59.

English Translation **Sauermann, Heinz and Reinhard Selten.** 1960. "An Experiment in Oligopoly," *General Systems, Yearbook of the Society for General Systems Research*. Ann Arbor, MI: Society for General Systems, 85–114, 206.

Reinhard Selten Yeah. It was unusual, of course, because experimental economics did not exist as a field. We made a search of papers before our paper appeared. There were twenty papers in the literature, mostly in the psychological literature beginning with *Thurstone*.[19] But in the '50s, it began to get more frequent but also there are very few before that. I think that experimental economics as a field emerged only in the '60s.[20] I don't know when. And though it was something very unusual, my fellow assistants would laugh at me and call me Doctor Mabuse, the gambler.[21] [some laughter]

And when there was a[n election] campaign of [the German Chancellor Konrad] Adenauer, which said "no experiments,"[22] they also showed it to me and said, "No experiments." So this were [unfinished]

Chris Starmer But it didn't worry you about crossing these boundaries?

Reinhard Selten It didn't worry me at all. We always sent, in our early experiments, discussion papers around. I must say that some people had doubts about the external validity of that,[23] but usually people did not oppose this. They found it quite interesting. It is not that these things got strong opposition. Of course, they were not dangerous for anybody, they were just a peculiarity.

Chris Starmer That is a subject I would like to come back to later on how results were perceived by and received by the relevant communities. I will definitely come back to that. I wonder now if I could just move to Vernon and ask you about your

[19] **Thurstone, Leon L.** 1931. "The Indifference Function." *Journal of Social Psychology*, 2, 139–67.

[20] **Editor:** Were you aware of agricultural experiments or time and motion studies? **Reinhard Selten:** "I knew of agricultural studies and time studies but I did not think of them as part of experimental economics. I understood experimental economics as aiming at the advancement of economic theory by the observation and analysis of the behavior of economically motivated subjects in laboratory or field experiments. I still think that this is a reasonable definition. It seemed to me, that the agricultural experiments and time studies, I knew of, did not contribute to the advancement of economic theory."

[21] The character Dr. Mabuse, a supervillain and a master of hypnosis, was introduced first in a novel *Dr. Mabuse, the Gambler* by Norbert Jacques in 1919 and featured in several popular movies.

[22] The slogan of 1961 campaign of the incumbent German government party was "Auch heute keine Experimente, CDU" [No experiments today either, the Christian Democratic Union].

[23] **Editor:** Were the doubts about experiments you mention explicitly phrased in terms of external validity at that time? **Reinhard Selten:** "We got letters from people to whom we sent discussion papers which expressed such doubts. However, they did not use the term "external validity." I do not remember when I first encountered this term but it was certainly only much later."

initial meetings with other experimental economists. I think that you met with *Sidney Siegel* in Stanford in the early '60s, perhaps in 1961. When did you learn about the research that was going on in Germany by Selten and others?

Vernon Smith: I don't recall. I do remember, I think it was in the '60s, Reinhard was running some seminars in Germany. And I was invited to one or two of those, but I never made it.[24] And I don't recall when I first met Reinhard.

Chris Starmer: Did you know other experimentalists before meeting Sidney Siegel?

Vernon Smith: Well, by the early '60s, and well before that actually, I knew of the Mosteller and Nogee experiment.[25] I don't know when. I would have read that pretty early. And I knew of the work that was summarized by *Andreas* [*Ortmann*][26] like the Thurstone experiments. There was something on social indifference curves by Hart[27] in 1930s. I never read Morgenstern's book.[28] I had a copy of the *Decision Processes* book since it originally came out, so I was familiar with that. In fact, when I first started teaching experimental economics, a graduate seminar in 1963, we didn't have enough literature of our own to really make a course that we were developing at Purdue where I used the

[24] Smith was invited to the first, second, and third International Conference on Experimental Economics that took place in 1971, 1977, and 1982 respectively as evidenced in **Smith, Vernon L.** "Smith Papers," *Vernon Lomax Smith Papers*. David M. Rubenstein Rare Book and Manuscript Library, Duke University, Durham, North Carolina.

[25] **Editor**: Were you aware of agricultural experiments, field experiments such as the negative income tax? **Vernon Smith**: "I knew of the agricultural production experiments designed to measure input substitution in the production of crops, milk, and so on, and it always had seemed to me the natural way to approach supply as well as demand issues. An important paper that I read and assigned in class was **Heady, Earl O.** 1957. "An Econometric Investigation of the Technology of Agricultural Production Functions." *Econometrica*, 25(2), 249–68. This paper would have appeared a year and a half after I started doing experiments in January 1956. Iowa State University had been an important center for this development."

[26] This is the paper that was sent to the participants in advance of the seminar to stimulate their memories.

[27] **Rousseas, Stephen W. and Albert G. Hart.** 1951. "Experimental Verification of a Composite Indifference Map." *The Journal of Political Economy*, 59(4), 288–318.

[28] Morgenstern's book was mentioned in Ortmann's paper. **Morgenstern, Oskar.** 1963. *On the Accuracy of Economic Observations*. Princeton, N.J.: Princeton University Press. (revision of the first edition from 1950).

Fouraker and Siegel, the working paper volumes or copies of those.[29]

[Besides the work outlined by Andreas] there was work in experimental games by social psychologists. Oh, and Ward Edwards' 1954 paper.[30] In fact, I met *Ward Edwards* early.[31] It would have been certainly by the early '60s. In the '60s at Purdue we regularly had visiting speakers in, and Ward Edwards would have come at least once. And also *Anatol Rapoport* who was at Michigan. In fact, I used Ward's work in decision-making under uncertainty, many of his papers in the early classes that I taught.

Now, as the literature began to develop in experimental economics, I replaced some of that early work. It became less of a course on decision-making and more of a course on markets. And I started to use some of the papers or books. Of course, *Bargaining Behavior* was published in 1963. Oh, and then there is the 1960 book. [inaudible] What's that?

Jim Friedman *Bargaining and Group Decision Making*.[32]

Vernon Smith Yes, I used that. But this was after I had met Sid Siegel.

[29] **Editor**: What other literature did you use? **Vernon Smith**: "The working papers included other authors, prominently *Martin Shubik* who knew and worked with *Sid Siegel*, and Sid's student Don Harnett. The final Fouraker Siegel book (1963) was published after Sid's untimely death at age 45 in 1962; it incorporated the work of Martin and Don from the working papers. Martin had worked with Sid and Larry *Fouraker* on the Cournot oligopoly experiments. *Siegel* and Harnett on the experiments using GE executives as subjects in their replication of the bargaining experiments. All are credited in the preface by Larry, but none survived as co-authors which I thought was ungenerous having used all the material in classes before it was published. Martin and Sid had a large agenda for further work that would have been path breaking at the time and established experimental economics much more prominently in the 1960s and 70s, but ended with Sid's death. People today have no idea of Sid's energy and depth."

[30] **Edwards, Ward.** 1954. "Variance Preferences in Gambling." *The American Journal of Psychology*, 67(3), 441–52.

[31] **Editor**: I wonder whether you interacted with *Ward Edwards* during your stay at USC in the academic year 1974/75. **Vernon Smith**: "Yes, I did. We got together sometime in that period, and I attended probably 2–3 of his Behavior Decision Theory conferences over in the valley in the 1970s, probably then, but also after I went to Arizona. I have long thought that the important early contributions of *Edwards* (whose father was a known economist) and *Anatol Rapoport* deserved more recognition. They were the pioneers that trained and created the generation of psychologists who (e.g. Slovic, Lichtenstein) did what was to be called Behavioral Economics."

[32] **Siegel, Sidney and Lawrence E. Fouraker.** 1960. *Bargaining and Group Decision Making; Experiments in Bilateral Monopoly*. New York: McGraw-Hill.

## Early Meetings and Seminars

| | |
|---|---|
| Chris Starmer | And I think in the early '60s, perhaps '63, '64, there were meetings at Carnegie Institute that you participated in. Is that right? You met other experimentalists there? |
| Vernon Smith | Oh, are you talking about the Ford Foundation? |
| Chris Starmer: | Yes, Ford Foundation Research. |
| Vernon Smith | *Lester Lave*[33] was at Carnegie Institute of Technology before they changed the name,[34] and Lester had done a couple or three papers on prisoner's dilemma games.[35] And we had been in communication.[36] *Dick Cyert*[I] at Carnegie Tech, also *Herb Simon* I'm sure would have been a factor in encouraging this, led us to make a proposal to the Ford Foundation to do summer faculty research workshops. The first one we did was in the summer of 1963. Then we did one in the summer of 1964. And as I recall, [Roger] *Sherman* was there. Jim, were you at either of those? |
| Jim Friedman | No. |
| Vernon Smith | Okay. |
| Jim Friedman: | I was at something—no that was later. I was at something at Berkeley, but that was [in] '68.[37] |
| Vernon Smith | Yes. And I don't know who else might have been at that. Oh, *Bill Starbuck* probably would have been at one of those. *Bill Starbuck* came to Purdue from Carnegie Tech sometime in the '60s. |
| Chris Starmer | Was this meeting focused on experimental economics? |
| Vernon Smith | *Bill Starbuck*'s focus was on everything. [laughter] He had wide-ranging interests. And if you were doing experiments, he was interested in that. He clearly had some exposure to |

[33] **Frank Trenery Dolbear Jr.** received his Ph.D. in Economics from Yale in 1963. He spent the next three years at Carnegie Institute of Technology. Since 1968 he has been at Brandeis University. He was active in experimental research only in the 1960s.

[34] Carnegie Institute of Technology became Carnegie Mellon University in 1967.

[35] **Dolbear, F. Trenery and Lester B. Lave.** 1966. "Risk Orientation as a Predictor in the Prisoner's Dilemma." *The Journal of Conflict Resolution*, 10(4), 506–15, **Lave, Lester B.** 1962. "An Empirical Approach to the Prisoners' Dilemma Game." *The Quarterly Journal of Economics*, 76(3), 424–36, **____.** 1960. *An Empirical Description of the Prisoner's Dilemma Game*. Santa Monica, Calif.: Rand Corporation, **____.** 1965. "Factors Affecting Co-Operation in the Prisoner's Dilemma." *Syst. Res. Behavioral Science*, 10(1), 26–38.

[36] See for instance correspondence from 1965 and Smith evaluation of Lave's experimental work from 1969. **Smith, Vernon L.** "Smith Papers," *Vernon Lomax Smith Papers*. David M. Rubenstein Rare Book and Manuscript Library, Duke University, Durham, North Carolina.

[37] In July 1968 *Austin Hoggatt* together with John T. Wheeler directed the Workshop in Experimentation in Management Science at University of California, Berkeley. Friedman attended for one month. See for instance, **Hughes, G. David and Philippe A. Naert.** 1970. "A Computer-Controlled Experiment in Consumer Behavior." *The Journal of Business*, 43(3), 354–72.

experiments and certainly, the idea of experimentation by the time he came to Purdue. He had been influenced by *Herb Simon* and the crew at Carnegie Tech. I am pretty sure that he probably was at that seminar. Later, he was instrumental in designing the first lab at Purdue.[38]

Chris Starmer: And would you think of that as a gathering of experimental economists? I suppose what I am getting at is was this one of the first meetings of experimental economics? Would that be a characterization?

Vernon Smith: Yes. And we went through the literature that was available at the time in 1963. And the participants were encouraged to work on projects and to do some experimental work.

## Chance Encounters and Conversions

Chris Starmer: Okay. Thank you. I think what I would really like to do now is to open things up to you to choose when you want to contribute and what you want to say. When I introduce a topic, if you want to speak to it, can I remind you of the convention to raise your hand. If there are many hands, I will try and keep a record of the sequence. I will try and acknowledge you when you have raised your hand, so you won't have to keep it in the air. You will hopefully know you have made my list. And it is one hand for a new theme and two if you have got some interjection, which is very closely related to what is being discussed as the present topic.

I should also say that if there are particularly popular topics that many people want to contribute to across the course of these sessions, it may not be possible to practically let everybody speak to every topic that they might have something to contribute to. But hopefully, over the course of the two days, there will be good opportunities for all to speak. But apologies if I, at times, move from topics before you have

[38] In an autobiographical essay *Starbuck* details his experience with experimental work in economics and psychology, influence of *Simon* and *Cyert* and his involvement in designing computerized social science laboratories (1967): **Starbuck, William H.** 1993. *"'Watch Where You Step!' Or Indiana Starbuck Amid the Perils of Academe (Rated PG)," Management Laureates: A Collection of Autobiographical Essays. Greenwich, Connecticut; London, England: JAI Press, 63–110.* See also **Cyert, R. M.; J. G. March and W. H. Starbuck.** 1961. "Two Experiments on Bias and Conflict in Organizational Estimation." *Management Science*, 7(3), 254–64. **Starbuck, William H. and Frank M. Bass.** 1967. "An Experimental Study of Risk-Taking and the Value of Information in a New Product Context." *The Journal of Business*, 40(2), 155–65. For the joint work with Vernon Smith on a laboratory at Purdue see Chapter 6, Footnote 8.

an opportunity to speak. I think it is inevitable that people will have more to say than we have a chance to hear.

As a first theme, I'm interested in exploring how people got started in doing experimental research. Perhaps I might ask you to think about two dimensions of this. One is how did you first become aware of experimental research? And secondly, what drew you into doing it? And so if I could invite people who would perhaps like to speak on that topic? Betsy.

Betsy Hoffman — Actually, I might cede to Charlie [Plott]. Charlie, did you have your hand up?

Charlie Plott — Yeah. I was going to say something about that.

Betsy Hoffman — Why don't you start, and then I will follow you because what I have to say follows from what Charlie is going to say.

Charlie Plott — At least my experiments, my exposure came through chatting with Vernon. I was at Purdue in the late '60s.[39] And Vernon continued to tell me about the convergence in his demand and supply experiments. I thought that his results were silly, and it was clearly not demand and supply, but it must have been a Bayesian game. So I had been touched by *Harsanyi*. I thought that I could build up priors in learning to get these systems to converge away from the competitive equilibrium and, therefore, show that it was not the law of supply and demand that was working but, in fact, it was a Bayesian game, which I considered to be quite different principles. I commandeered a graduate student named *Harvey Reed* who did this [experiment] and immediately he demonstrated that my beliefs were correct.[40, II] Later, I found out,[41] however, that *Harvey* was a terrible experimentalist, and the procedures he was using were really embarrassing.[42]

---

[39] Plott was at Purdue from 1965 until 1971.

[40] **Reed, Harvey Jay.** 1973. "An Experimental Study of Equilibrium in a Competitive Market," Purdue University, Plott and Carl H. Castore were on Reed's committee.

[41] Paper presented "An Unsuccessful Attempt to Experimentally Discredit the Law of Supply and Demand" (with H. Reed), Western Economics Association Meeting, Las Vegas, 1974.

**Editor:** When did you find this out? **Charlie Plott**: "I found that out in 1973 when I had a student attempt to replicate Harvey's experiments. Much later I published a paper that discussed the matter and introduced the concept of "reparameterization" as a way to understand and interpret the data as actually supporting the theory that the experiments were originally intended to explore."

[42] **Editor**: Why were Reed's procedures embarrassing? **Charlie Plott**: "The procedures he used are now known to create market inefficiencies. He was conducting a market experiment following Vernon's work when I was still at Purdue, probably somewhere near 1970. Harvey had an environment with two units and the appropriate way to induce the incentives was not developed until my work with Fiorina and the proper way to conduct the double auction when individuals could trade multiple units was not developed until my much later work with Vernon."

The next encounter [with experiments] was again shadowed with Vernon. This is around 1969, 1970,[43] and I was interested in the mathematics of axiomatic social choice theory and voting. These are public goods environments. And I realized that one could take Vernon's idea about induced preferences and induce them in a much broader economic environment. In that sense I began to test things that were evolving out of voting theory and out of cooperative game theory without side payments, which is much different from the bargaining problem.[III]

We were studying the core, bargaining sets and [were] doing so within institutions that could be precisely defined. Features of Robert's Rules[44] then led us into discovering that there was a host of principles coming out of game theory, not cooperative game theory, not with games with side payments, but out of dominance relation and treating these abstractly that were extremely powerful in demonstrating and predicting. And I became associated with *Morris Fiorina* who had worked with *Bill Riker*, who was essentially doing the same thing in political science.[45, IV] Bill and Mo had studied many, many procedures.[V]

That is how I got started. Then Vernon came to Caltech where he began to get more focused on markets and market institutions.[46] That set that stage[VI] and, of course, there is many, many, many results and discoveries that evolved from there.[VII]

Chris Starmer: Thank you. Betsy, do you want to follow?

Betsy Hoffman: Yes, I was pretty sure what Charlie was going to say, and I knew that I would follow directly on from what he was saying. I came to Caltech in 1975 as a graduate student. Having been a historian, I have a Ph.D. in history as well,[47] I came to Caltech recruited by *Lance Davis* for the sole purpose of improving my quantitative and theory skills so that I could go back to economic history with a new set of tools. I came to Caltech with a very

[43] The exact dates remain unclear, but dating to 1970–1971 seems as more precise.

[44] Robert's Rules of Order is a set of rules for running meetings and conferences.

[45] After moving to Caltech in 1971.

[46] Smith visited Caltech, where he received his undergraduate degree in 1949, as a Fairchild Scholar in the academic year 1973–74. Bill Riker was also in residence.

[47] Hoffman earned her Ph.D. in history from University of Pennsylvania in 1972. The dissertation was entitled "The Sources of Mortality Changes in Italy since Unification." It later appeared in book form **Hoffman, Elizabeth.** 1981. *The Sources of Mortality Changes in Italy since Unification*. New York: Arno Press.

clear purpose. In fact, I didn't even intend to finish a second Ph.D. I really came just to get the tools. It was like a post-doc.

But after my first year at Caltech, I got talking to an economist who told me that the jobs were much better in economics, and this was somebody outside of Caltech.[VIII] This was somebody who didn't have a particular personal interest in my continuing in economics and said that I would really, as long as I had invested as much time as I had and I had taken all the classes and I passed prelims, that there was huge benefit to my finishing the Ph.D. in economics, even if I went into economics as an economic historian. I continued [with the doctoral program], and I actually ended up writing a dissertation that is a quantitative economic history dissertation on the Colorado River compact.[48]

I wrote a dissertation that is totally unrelated to experimental economics and much more related to the purpose for which I went to Caltech. But my last quarter there,[49] I had to take some classes, and I had taken all the classes that I felt were important for the purpose for which I had come. But as long as I was going to finish a Ph.D., I had one more quarter of classes I had to take. *John Ferejohn* was my mentor, my advisor, and I went to him and I said, "So John, what's left that I should take?" And he said, "Well, you can't leave Caltech without taking Charlie's seminar in experimental economics." And I said, "Why? What good is it going to be to me as an economic historian?"

And he said, "It's just that this is the most important thing that is happening at Caltech right now," and "you can't leave Caltech without taking the seminar." I said, "Fine." I mean, why not. I had three courses I had to take to check off the last three boxes. And it changed my life. I took this seminar, and it was the most fun. I had lots of fun as an economic historian. I had done lots of fun things. But this was the most fun I had ever had.

Chris Starmer Okay. I'm going to make a mental note of the fact that it changed your life, and I'm going to return to that when we come to the section on skills and ask you about that again.

Betsy Hoffman That is fine.

Chris Starmer Al, you signaled a little while ago.

---

[48] **Hoffman, Elizabeth.** 1979. "Essays in Optimal Resource Allocation under Uncertainty with Capacity Constraints," Pasadena: California Institute of Technology.

[49] **Editor**: The chronology does not seem correct. **Betsy Hoffman**: "It was the third quarter of my second year. I wrote my dissertation the next year and graduated in 3 years."

Al Roth: I came from a different tradition. I studied game theory, and I took a game theory course when I was a graduate student at Stanford[50] from *Michael Maschler* who was visiting from the Hebrew University. He didn't talk much about experiments, but he talked a little bit about the work of *Amnon Rapoport* who was interested in the bargaining set, which Michael was interested in. There were lots of, not lots, but there were experiments in game theory going back a way starting with the prisoner's dilemma maybe.[51] When I came to the University of Illinois,[52] having written a theoretical thesis, one of the other fellows who came just as I did was *Keith Murnighan*, a social psychologist from Purdue.

He and I got along well, and we thought that we would try to do some experiments on the games that I had written my dissertation about. For him, it was natural to go into the lab, and there were lots of experiments by psychologists on bargaining and on group interactions and things like that. That was the literature I was initially aware of. I think probably the first experiments by someone around this table that I became aware of were Charlie's committee experiments having to do with the core,[53, IX] because, again, those were in the game theoretic tradition.

Chris Starmer: John and Reinhard?

Reinhard Selten: I wanted you to add that *Mike Maschler* was actually one of the first experimenters. He did experiments maybe around 1960 with cooperative characteristic function games. Before he got a university position, he was a high school teacher. He had his high school class play characteristic function games with 3 or 4 players. They bargained face to face about coalitions to be formed and the division of payoffs among the members. There was no time restriction. At the end of a game they had to turn in a card showing what they had agreed

[50] Roth studied at Stanford between 1971 and 1974.

[51] **Kalisch, Gerhard K.; J. W. Millnor; John F. Nash and E. D. Nering.** 1954. "Some Experimental N-Person Games," R. M. Thrall, C. H. Coombs and R. L. Davis, *Decision Processes*. New York: Wiley, 513–18.

[52] Roth moved to the University of Illinois in 1974 and remained until 1982.

[53] This is the already mentioned seminal paper **Fiorina, Morris P. and Charles R. Plott.** 1978. "Committee Decisions under Majority Rule: An Experimental Study." *The American Political Science Review*, 72(2), 575–98. and **Isaac, Mark R. and Charles R. Plott.** 1978. "Cooperative Game Models of the Influence of the Closed Rule in Three Person, Majority-Rule Committees: Theory and Experiment," P. Ordeshook, *Game Theory and Political Science*. New York: NUY Press, NA.

upon and for each player separately the reasons for her or his conduct.

I met him at Princeton in '61 at a game theory conference[54] organized by *Oscar Morgenstern* who had invited me and made it possible for me to be there. At this conference Mike Maschler already had a list of all his experimental results with him. He must have run his experiments in 1960 or even before. Later he sent me his research report.[55] He had submitted it for publication to the *Journal of Conflict Resolution* and they asked him to shorten it, but he never did this.

At one of the conferences in Germany on *Bargaining* and *Coalition Formation* I told him that we would in our volume print his research paper completely as it was.[X] He wouldn't have to change anything. We will just print it. So almost 18 years later, it was published.[56]

*Maschler's* research report had a great influence on my own work. I very carefully looked at his experimental results and the reasons given by the players. Finally I came up with something called "equal share analysis," a behavioral theory about n-person games in characteristic function games. His work was very important for me.

Chris Starmer: I will take perhaps three more on this point. John, Charlie, and Jim. John.

John Kagel: I came to this in a totally different way. I was at Purdue when Charlie was on my comprehensive exam committee; I remember distinctly failing his question. But the interesting or perhaps surprising thing is that there was no history of Vernon's work since Vernon had left by the time I arrived.[57] There was no history of that work there or any of that being

---

[54] Published privately in 1962 as **Princeton University, Conference.** 1961. "Recent Advances in Game Theory; Papers Delivered at a Meeting of the Princeton University Conference, October 4–6, 1961," *Recent advances in game theory*. Princeton, N.J.: Princeton University Press. The introduction was written by Maschler. Contributions were made by Morgenstern, Vickrey, Fouraker, Suppes, Afriat, Aumann, Shapley and many others.

[55] **Maschler, Michael.** 1972. "Equal Share Analysis of Characteristic Function Experiments," H. Sauermann, *Contributions to Experimental Economics = Beiträge zur Experimentellen Wirtschaftsforschung*. Tübingen: Mohr, 130–65.

It was based on an 1965 working paper ____. 1965. *Playing an N-Person Game, an Experiment*. Princeton, N.J.: Princeton University, Econometric Research Program Research Memorandum No. 73.

[56] ____. 1978. "Playing an N-Person Game: An Experiment," H. Sauermann, *Bargaining Behavior. Contributions to Experimental Economics = Beiträge zur Experimentellen Wirtschaftsforschung*. Tübingen: Mohr, 231–328.

[57] Kagel started at Purdue in 1967 and he received his doctorate in 1970.

taught.[XI] I got into doing experiments because I was a graduate student of *Bob Basmann*. And *Basmann* was an econometrician who was very concerned with that the data correspond to the primitives in our economic models. And he was also interested in individual choice.[XII]

The standard field data that is available for that kind of thing just didn't fit. In terms of looking at that I had come across the experiments by *Thurstone* and May[58] and those people. *Ray Battalio* and I were just looking for a place where we could collect individual consumer choice data. It was just by happenstance that we learned about token economies.[59] And by happenstance there was a gentleman [Robin C. Winkler] at Stony Brook at the time who was a psychologist and was also interested in economics of token economies. There was a natural connection there. We hooked up and started to do our first experiment in a backwater of a mental institution, which is to say the least somewhat unusual.[60]

At one point, you raised the issue of how were results received by different communities. Well, you can imagine how some of these results were received by certain communities. One of the interesting elements of how it was received by one community though was at the time that we were doing the experiment, I gave a talk to the psychology department at Stony Brook. I don't think anyone in economics bothered to show up. They wouldn't have advertised it there. We were talking about revealed preference theory and things like that.

One guy got up in the middle of the talk and said that if economics is so primitive, I have got nothing to learn here and walked out of the room. [some laughter] At the end of the talk, *Howard Rachlin* and *Leonard Green* came up to me and said, "We've already done this sort of experiment." I said,

---

[58] **May, Kenneth O.** 1954. "Intransitivity, Utility, and the Aggregation of Preference Patterns." *Econometrica*, 22(1), 1–13.

[59] **Ayllon, Teodoro and Nathan H. Azrin.** 1968. *The Token Economy; a Motivational System for Therapy and Rehabilitation*. New York: Appleton-Century-Crofts, **Kagel, John H.** 1972. "Token Economies and Experimental Economics." *Journal of Political Economy*, 80(4), 779–85.

[60] **Battalio, Raymond C.; Edwin B. Fisher; John H. Kagel; Robert L. Basmann; Robin C. Winkler and Leonard Krasner.** 1974. "An Experimental Investigation of Consumer Behavior in a Controlled Environment." *Journal of Consumer Research*, 1(2), **Battalio, Raymond C.; John H. Kagel; Robin C. Winkler; Edwin B. Fisher; Robert L. Basmann and Leonard Krasner.** 1973. "A Test of Consumer Demand Theory Using Observations of Individual Consumer Purchases." *Economic Inquiry*, 11(4), 411–28.

"Yeah? Okay, show me." And that led to our designing our own experiments.[61] I learned about these other people when [*Martin*] *Shubik* put a session together at the AEA meetings. I don't know what year that was.[62] So just my ignorance about what other people were doing. And we found out about Vernon and Charlie and other people.

We immediately wrote to them and told them what we were doing.[63] There wasn't email back then, so it took a little while. But that is how I became aware of what other people were doing.

**Chris Starmer** Thank you. Is that a one hand up Vernon? It's a two hand up.

**Vernon Smith** I think it is great that John started on this [experimental work] uncontaminated by anyone else because this was, I think, very exciting work that he and *Ray* [*Battalio*] did. In fact, it was the only game in town at the individual decision and classic preference theory work. There is one thing I wanted to mention that is in the *Ortmann* paper. He refers to the Radford paper on economics of a prisoner of war camp.[64] I learned of that work very early. In fact, it was pretty widely known, and a lot of economists were interested in that paper. It was just a really wonderful contribution.[XIII] It helped to influence me on the possibility of experimental [work]—although that was not a controlled experiment.

**Chris Starmer** Thank you. Charlie.

**Charlie Plott** Just in terms of a couple of names. *Keith Murnighan*, as it turns out, was at Purdue. *Keith* was a student of *Carl Castore* who was a psychologist, and *Keith* and *Carl* sat in my social choice course.[65] We actually looked at voters' paradoxes. Now, the interesting thing about that was that *Carl* listened to this, and he saw the cycle in voting,[66] and he says we can test that. He decided to get [LP long play vinyl] records, [and follow] the

---

[61] **Kagel, John H.; Raymond C. Battalio; Howard Rachlin; Leonard Green; Robert L. Basmann and W. R. Klemm.** 1975. "Experimental Studies of Consumer Demand Behavior Using Laboratory Animals." Ibid.13(1), 22–38.

[62] This could be the 1975 AEA session which was, however, not organized by Shubik but by Gary Becker.

[63] The earliest letter from Charlie Plott to John Kagel is from January 1977, and Battalio from June 1974. **Plott, Charles R.** "Plott Papers," *personal archive of Charles Plott.* California Institute of Technology,

[64] **Radford, R. A.** 1945. "The Economic Organization of P.O.W. Camps." *Economica*, 12, 189–201.

[65] That was in the academic year 1970–71, just before Plott left for Caltech.

[66] The voting paradox or Condorcet's paradox is a situation in which collective preferences can be cyclic (i.e. not transitive), even if the preferences of individual voters are not.

typical psychologist approach. He went around and asked people their preferences for these big records. And so then he was going to have a group choose a record that everyone could get and structured them so there was intransitivity.

The interesting thing about it was that at that time, it wasn't clear what he was going to learn. We didn't realize that procedures [were overwhelmingly important]. We didn't have the dominance relation looking at it. We really hadn't separated the idea of games without side payments [from games in characteristic function form], which we did later.[67] And so he did this [experiment], and it seemed he didn't learn anything. But that was a clear step towards the problem of trying to control preferences in these very complex areas where everything is of public goods [nature].[XIV] *Carl Castore* in some sense was an instrumental step. He even got some money to study this. I think that nothing finally came of it, but he is a very interesting name.[68]

And another name that is interesting with respect to Purdue was a guy named *Cliff Lloyd*. *Cliff Lloyd* was absolutely fascinated with the problem of testing preference theory. He thought preference theory was testable. And *Cliff* had an influence; you could see a lot of heads shaking here.[XV] *Cliff* had an influence on a lot of us because he looked to the second order conditions and the symmetry of the substitution matrix, and he said that substitution matrix is just loaded with testable propositions. He spent much of his time trying to understand how one would set up an experiment. He even tried to contract for a little village in Alaska so he could actually control the incomes to test for the symmetry of the substitution matrix.[69]

---

[67] Reference to non-cooperative game theory. See earlier statement by Plott. For a discussion of the dominance relation see **Plott, Charles R.** 1976. "Axiomatic Social Choice Theory: An Overview and Interpretation." *American Journal of Political Science*, 20(3), 511–96.

[68] In fact, Castore and Murnighan published a study based on their LP experiments. However, they did not investigate intransitivities to any extent. **Castore, Carl H. and J. Keith Murnighan.** 1978. "Determinants of Support for Group Decisions." *Organizational Behavior and Human Performance*, 22(1), 75–92.

Earlier versions appeared in 1972–3 as working papers **Castore, Carl H.** 1972. *Intragroup Concordance and the Effectiveness of Majority Rule Decisions*. Ft. Belvoir: Defense Technical Information Center. Purdue Univ Lafayette, **Castore, Carl H. and J. Keith Murnighan.** 1973. *Decision Rule and Intragroup Goal Concordance as Determinants of Individual Reactions to and Group Decisions*. Ft. Belvoir: Defense Technical Information Center. Purdue Univ Lafayette.

[69] The location of Lloyd's experiment was Postville, Newfoundland, in Canada, but he considered other locations as well. This research was published only posthumously under the title *Northern Store Project*. **Lloyd, Cliff.** 1980. *The Collected Works of Cliff L. Lloyd*. Burnaby, B.C.: School of Business Administration and Economics, Simon Fraser University.

Anyway, that was *Cliff's* major thing. He was also a student of *John Hicks*. There is a real continuity of this kind of asking what are the principles of individual decision-making that might lead to testable propositions. Anyway, what influenced all of us at Purdue was *Cliff's* worrying about this particular problem [of empirical support of theories]. He was leading all of us to ask such questions about these complex theories. What features of models might lead to a testable proposition? The idea of an experiment was never in question. The idea of a testable proposition was never in question. All this was just a second nature for most of us who were in that environment.

Now, in terms of receptiveness, let me make one comment. In the late '60s I moved away from individuals into groups [decision making]. I think that was a major transition.[XVI] When we started then going to the [professional] meetings and submitting our papers, they organized us in economics of education.[70] [some laughter] Namely, they looked at all of our work as nothing but pedagogical devices. There was no science there [for the organizers] at all. [XVII]

## The JEL-Code and Closeted Experimentalists on the Job Market

| | |
|---|---|
| Chris Starmer | How do you think you should have been classified if you had been able to choose? |
| Charlie Plott | How should we have been? |
| Chris Starmer | Yes. |
| Charlie Plott | Well, I think that we were dealing with the [empirical] foundations of economics [but] economics does not have [such] a classification. If the experiment was a committee experiment, I would have put it having to do with something with public choice. If it was a market experiment, I would have had it in microeconomics. I wouldn't have separated it out as anything special. It is data about phenomena [and the |

[70] **Editor:** I have not found any evidence for this so far. **Charlie Plott:** "The first experimental paper I gave at a professional meeting was placed in a section of education. It might be noted that in the 1960's and perhaps today, there was a healthy use of hands on methods to demonstrate economics but it was not viewed as experiments. Some of the economists who were focused on economics education were at Purdue."

empirical relationships the data present]. But that is the way it was treated—just education.

Chris Starmer: Thanks, Charlie. Betsy.

Betsy Hoffman: I wanted to build on that, and this is 10 years later. I'm on the job market in 1978 and was basically counseled, even by Charlie, not to talk about this really exciting work I was doing in experimental economics,[XVIII] but to focus in my job talks completely on my work on the Colorado River compact and to never mention experimental economics. John [Ledyard] attended one.

Chris Starmer: And why was that?

Betsy Hoffman: Because it was considered that I would be a pariah on the job market if I sold myself as an experimental economist. But the junior faculty, at every university where I went to give a talk, the junior faculty would whisk me into their offices and close the door and say okay, I want to hear about experimental economics.[71]

Chris Starmer: And were you starting to think of yourself as an experimental economist at that point?

Betsy Hoffman: I was in both. I had my feet in both camps [economic history and experimental economics]. John [Ledyard] really helped me get the interview and get the job at Northwestern.[72] I know that even though he never said anything about it. But it was the economic historians who took me under their wing and really probably made me get hired. Now, John greased the skids, but I think without *John Hughes* and *Joel Mokyr*, it would have been hard to persuade the rest of the department to hire me. The economic historians adopted me. I taught economic history. I was part of the economic history seminar. I was writing and published a book in economic history.[73]

But I was starting to do experiments, and I started my work with *Matt Spitzer* who had been at Caltech with me and was in the law school [at Northwestern]. I started my work with *Ed Packel* who was and still is at Lake Forest. Matthew and I put in the first NSF grant, and I know we are going to be talking about

[71] **Editor**: Where did you have your interviews? **Betsy Hoffman**: "I interviewed at Arizona, Swarthmore, Washington, Iowa, Ohio State, Northwestern, and Boston College. I had more scheduled, but cancelled the rest when I got offers from Northwestern and Swarthmore, my top choices."

[72] Ledyard was a Fairchild scholar at Caltech in the academic year 1977/78.

[73] **Hoffman, Elizabeth.** 1981. *The Sources of Mortality Changes in Italy since Unification.* New York: Arno Press.

funding later. But it was very clearly, even in 1978 and 1979, not a respected thing to do.

Chris Starmer: Charlie.

Charlie Plott: I have a comment on that. It is true, these students that came out in political science and economics, even though they were fire-breathing experimentalists, we counseled them to actually sell themselves as something else. "Don't sell yourself as an experimental economist. There are no jobs for experimental economists. There are jobs for traditional people." Say, "I do the traditional stuff, and oh by the way, I also have an interest over here." But, approach experiments strictly as an aside thing.

But then once the camel has his nose under the tent, then it can expose itself as something else, and that seemed to work quite well. But when getting in the door, there was a solid block.

Betsy Hoffman: But when I came up for my third year review,[74] I got the explicit recommendation of—stop doing experimental economics. You are never "going to become famous fast enough." I will never forget that quote. Go back to doing economic history.

John Ledyard: Who was it?

Frans v. Winden: That is a good question

Charlie Holt: For the record.

Betsy Hoffman: Start naming names and I will –

John Ledyard: *Leon Moses*?

Betsy Hoffman: Who?

John Ledyard: *Moses* wasn't an economic historian.

Betsy Hoffman: No. He was a theorist. He was chair of the department at the time.

John Ledyard: Was that *Dale Mortensen*?

Betsy Hoffman: No, it wasn't *Mortensen*.[75]

John Ledyard: I don't remember.

Betsy Hoffman: Anyway, I was actually recruited to Purdue to be an experimental economist, so that was how I launched a career as an experimentalist after being told that I was not going to get famous fast enough, so that would have been in 1981.[XIX] I was recruited to Purdue, and I haven't done an economic history paper since.

Chris Starmer: John and then Steve.

[74] Hoffman came up for a third year review in the fall of 1981.

[75] **Editor**: Who was it then? **Betsy Hoffman**: "Actually it was *Dale Mortensen*."

John Kagel — I just wanted to make a quick follow up to what Charlie said about how these things should have been classified. I agree 100 percent that they should be classified by the topic, by the subject matter of whether you are dealing with, say, auctions or you are dealing with voting and this sort of thing, because it is a tool. It is not like econometrics. It is very far from econometrics where there are real high-powered techniques that are being developed all the time. I think it is more an approach to looking at questions. And it should be in the context of those questions. We should not be talking just to ourselves in terms of our professional work.

Stephen Rassenti — Well, my introduction to experimental economics was entirely different. I was an engineering graduate student at Arizona, and I remember hearing experimental economics or what Vernon was doing there as referred to as southwest economics disparagingly. I don't know where the expression came from. [some laughter] But Vernon introduced me to a couple of topics that I eventually used in my dissertation as an engineer. And I went off to work with Bell Labs after that and came back to experiments later, but it was not directly a part of my dissertation.[76, XX]

Chris Starmer — Jim, do you want to follow?

Jim Friedman — Actually, there are two things I would like to comment on. One of them popped up after I put my hand up before. The first harks back to your mentioning my dissertation perhaps being the first experimental one. There was another graduate student at Yale at the same time that I was there who was finishing an experimental dissertation just when I was. His name was *Trenery Dolbear*. [Smith: ehm][XXI] And you [John Ledyard] probably knew him at Carnegie.[77] *Tren* was in a cohort that came to graduate school a couple of years before I did, so I was acquainted with him, but I didn't know him very well. I wasn't aware that he was doing an experimental dissertation until we were both on the job market.

*Dick Cyert* who was the Dean of the GSIA at Carnegie had us both come out for job interviews. He did the unusual thing of bringing us out simultaneously, so we flew out and back together, stayed together, and then we both had job offers.

[76] **Rassenti, Stephen.** 1982. "Zero/One Decision Problems with Multiple Resource Constraints Algorithms and Applications," *Systems and Industrial Engineering*. University of Arizona.

[77] **Editor**: Did you know him, John? **John Ledyard:** "I was aware of Dolbear, but did not know him personally."

And I would like to comment on what you were saying, Betsy and you Charlie, about the admonitions about going forth as an experimentalist. In 1963, which I think was before 1981, [Hoffman: oh yes, considerably] the only arrow in my quiver was oligopoly experiments. I did not apologize for that and say I did this junk for my dissertation, but I will do something real later. When I was asked in an interview what my plans were, at that point, I had plans for further experiments.

I did have some good job offers at the time, including Carnegie where they were seriously interested in my experiments. Penn where they were not seriously interested in my experiments, I think, but they made me a job offer. And Yale where there was really nobody with any serious interest in experiments, where I stayed and continued in an environment in which nobody cared a lot about what I was doing but was very supportive that I should do it. So I have a little different take. Also, the first article I sent off to a journal was one of the easiest acceptances I ever had. I know I don't have to speak to this group of people about how editors and referees treat one's brilliant papers.

I have had my share of rough treatment, sometimes fair and sometimes unfair, but that experimental paper following my dissertation sailed into *Econometrica* in a way that gave me delusions about how easy it was.[78] [some laughter]

Chris Starmer Vernon.

## Institutional Settings

Vernon Smith Well, Purdue and *Carnegie Institute of Technology* were pretty different [from other places], [Hoffman: Yeah] and I think this tells you why it is that the first Ford Foundation summer fellowship was sponsored with Purdue and Carnegie. As a faculty member at Purdue,[79] I never felt the least bit [different.] In fact, I was encouraged to do what I was doing.

[78] **Friedman, James W.** 1967. "An Experimental Study of Cooperative Duopoly." *Econometrica*, 35(3/4), 379–97.

[79] Smith stayed at Purdue from 1955 until 1967 with a hiatus at Stanford during the academic year 1961/2.

It is just when I got outside of Purdue, I found the world much different. The Purdue program really—the faculty there, which consisted of *Stan Reiter*, *John Hughes*, later *Nathan Rosenberg*, *Lance Davis*, [Plott: *Jim Quirk*] *Jim Quirk*, *Saposnik*. For example, when we created the *School of Industrial Administration* in 1957, not all of those people were there yet. We decided not to have departments. And what you are hearing around this table is why we didn't want to have departments.

The Purdue program was built on basically three things: economic theory, quantitative methods, and economic history. The quantitative methods included econometrics and experimental economics. Basically, what was in the program was whatever the faculty was doing. It was a faculty that was geared to developing "knowledge-how" more than "knowledge-that." I think that this had a lot of similarities with what was going on in Pittsburgh. In fact, John [Ledyard], you went to Pittsburgh.

John Ledyard: I got lucky and went to both places.

Vernon Smith: Yeah. I think that really does a lot to explain that tolerance for doing unusual things [which] was an important part of the early development of experimental economics.

Jim Friedman: But that tolerance did exist at some of the more standard places as well.

Vernon Smith: Ehm.

Betsy Hoffman: In very small measures.

Chris Starmer: Charlie [Holt].

Charlie Holt: I was a graduate student at Carnegie Mellon.[80] The great thing about Carnegie was they encouraged the faculty to write papers with the graduate students. I worked on two papers with *Dick Cyert* who was president of the university then, and *Morris DeGroot* was a statistician and student of [*Leonard, Jimmie*] *Savage*.[81, XXII] Every Saturday, we would go into the president's office and we would sit down, and *Dick Cyert* would say—it was the behavioral economics tradition—this is the way decisions are really made in the world. For example, when we make investment decisions, we don't look at rates of return very carefully. We look at what are

[80] Charlie Holt studied at Carnegie Mellon from 1970 until 1977. During 1971–1973 he was stationed in Japan as a Navy Reserve.

[81] **Cyert, Richard Michael; Maurice H. DeGroot and Charles A. Holt.** 1979. "Capital Allocation within a Firm." *Syst. Res. Behavioral Science*, 24(5), 287–95, ____. 1978. "Sequential Investment Decisions with Bayesian Learning." *Management Science*, 24(7), 712–18.

the retained earnings? What do we have to work with? We went back and redid some of Jorgenson's work adding retained earnings.[82] I remember it was the feeling of here is the world, this is the way the business world works. And I got a lot out of that.

At the time, as graduate students, we were much more excited about *Lucas* and *Prescott* and rational expectations. And so I remember once, later when I was at Minnesota, and we hired *Ed Prescott* to come there. He told me, he said, "Charlie, you shouldn't do experimental economics. It was a dead end in the '60s and this could be a dead end in the '80s." I didn't listen to him.[XXIII]

Chris Starmer John.

John Ledyard Up until very recently, my life was as a theorist and not as an experimentalist. But I remember similar things happening to me when I went on the job market as a theoretical economist, as a mathematical economist. People didn't much like them either. Northwestern wouldn't interview, so it wasn't just anti experiments. Economics was growing up in this time, and there were a lot of different branches, many of which weren't particularly cherished by traditionalists. I want to second the thing Vernon said about Purdue and Carnegie and sort of Caltech follows in this model of no departments, and you are just working on ideas.

*John Hughes* once told me when I was a graduate student, he said one of my fellow graduate students complained that they were having to learn mathematics, and they complained to *John Hughes* who was an economic historian who they thought would lend a gentle ear to the statements. John looked at them and said economics is what economists do, and Purdue economics is what Purdue economists do, and you have to learn what we are doing and that's it.

Chris Starmer Quick two hander from Al.

Al Roth This question of resistance to experiments. In the '70s, *Keith* [*Murnighan*] and I mostly sent our papers to psychology journals.[83] But the game theory community was interested in

[82] **Jorgenson, Dale W. and Calvin D. Siebert.** 1968. "A Comparison of Alternative Theories of Corporate Investment Behavior." *The American Economic Review*, 58(4), 681–712.

[83] From the eleven joint publications in the period 1977 to 1988 (eight until 1983) only two were in psychology journals excluding two in the Journal of Conflict Resolution, which, according to **Al Roth,** "was in the late 1970s and early 1980s an interdisciplinary journal with political science flavor."

**Murnighan, J. Keith and Alvin E. Roth.** 1980. "Effects of Group Size and Communication Availability on Coalition Bargaining in a Veto Game." *Journal of Personality and Social*

experiments. There wasn't a lot of opposition to it. I'm reminded by Jim's early experience with referees, when I first started sending experiments to economics journals, it looked like life was going to be easy. *Keith* and I sent a paper to *Econometrica*, and I think that maybe they didn't even ask us to revise it.[84] The game theorists who reviewed it liked the idea that they were experiments. The referees' reports in those days didn't look like they do today where experimentally literate people look at it and talk about the experiment. Rather the referee's reports would say something like this paper reports an experiment. That's a nice idea. Let's publish it.

Chris Starmer Very last word, Frans.

Frans v. Winden I think many of these things escaped the minds of many people in Europe, except for the German speaking countries where there was already pretty soon an association in the mid '70s.[85] There was the very important early work by *Allais*.[86] But apart from that little was going on. Personally, I traveled a long road before I got to my own economic experiments and their design in the late '80s. I studied economics in my undergraduate studies and I did a minor in social psychology and a major in economic sociology. There I picked up some interesting work. I remember two textbooks that were very influential for me and really very interesting.

One was a very general introductory textbook in social psychology by Kretch, Crutchfield, and Ballachey *Individual in Society*.[87] The other one, which I liked even better, was by Cartwright and Zander.[88] It was on group dynamics and I was very much intrigued by that. Among the authors were

*Psychology*, 39(1), 92–103, **Roth, Alvin E. and J. Keith Murnighan.** 1978. "Equilibrium Behavior and Repeated Play of the Prisoner's Dilemma Games." *Journal of Mathematical Psychology*, 17(2), 189–98.

[84] Murnighan and Roth published together only one paper in Econometrica, which was submitted in May 1981 and revisions were received in November 1981. ____. 1982. "The Role of Information in Bargaining: An Experimental Study." *Econometrica: Journal of the Econometric Society*, 50(5).

[85] *Gesellschaft für experimentelle Wirtschaftsforschung* [Society for Experimental Economics Research] was founded by Heinz Sauermann in 1977.

[86] **Allais, M.** 1953. "Le Comportement De L'homme Rationnel Devant Le Risque: Critique Des Postulats Et Axiomes De L'ecole Americaine." *Econometrica: Journal of the Econometric Society*, 21(4), 503–46.

[87] **Krech, David; Richard S. Crutchfield and Egerton L. Ballachey.** 1962. *Individual in Society: A Textbook of Social Psychology*. Tokyo [etc.]: McGraw-Hill Kogokusha.

[88] **Cartwright, Dorwin and Alvin Frederick Zander.** 1960. *Group Dynamics: Research and Theory*. Evanston, Ill: Row, Peterson. 2nd edition appared in 1960 and the last in 1968.

Cartwright and Zander, Festinger, French and Lippitt. They formed a group around *Kurt Lewin*, who developed a field theory,[89] and they all worked at the University of Michigan. They tried to combine experimentation with formal mathematical modeling.

I don't think they were very successful in that respect, but I can remember that there were formalizations, and it was very interesting. I also must have heard about *Siegel* and *Fouraker*, and *Simon* in the context of economic sociology [and social psychology]. Later I got interested in political economy, [more specifically] the endogenization of political behavior in economic models. There I picked up an interest also in simulations from a more macroeconomic perspective. I heard about these simulation games or management games. Some of these were written for teaching actually. I learned about running macroeconomic models [on a computer] where you could steer variables like the interest rate and stuff like that.

I got interested in having a macroeconomic orientation with some micro underpinning. And in my Ph.D. thesis I did some simulations myself—numerical experiments, as I called them.[90] What really struck me at the time was the huge influence of econometricians, like *Tinbergen* and *Theil*, and the little interest they took in fundamental aspects of behavior, that is, what actually explains behavior; political behavior [in particular]. Through my background in social psychology and economic sociology, I had a big interest in this.

I'm pretty sure that I tracked what was going on. I came across the work by Charlie [Plott] on political decision-making, [like] voting and committee decision-making and things like that. I attended the public choice conference in San Francisco in 1980, and I met *Axelrod* there, for instance. I came [also] across experimental work. Then the key events for me were first that I visited Charlie [Plott] at Caltech where I was interested in the political economics that was developed

[89] Lewin proposed in his field theory that human behavior is a function of both the person and the environment.

[90] A slightly adjusted version of 1981 Ph.D.-thesis appeared **van Winden, Frans.** 1983. *On the Interaction between State and Private Sector : A Study in Political Economics*. Amsterdam; New York: North-Holland Pub. Co. ; Sole distributors for the U.S.A. and Canada, Elsevier Science Pub. Co.

there by people like Rod Kiewiet and *David Grether* and got also interested in, of course, experimental economics.[91]

I can remember that Charlie [Plott] and I talked about an experiment with overlapping generations. I was interested, for instance, in social security. How a pay-as-you-go system might be sustainable over rounds. We discussed it a little bit, but it never materialized. But I got the opportunity to participate in experiments and learned how to write instructions, do the designs, etc. That was '86, and then in '87/'88 [a second key event took place] when I visited the ZiF: *Zentrum für interdisciplinäre Forschung*, in Bielefeld, where Reinhard [Selten] was organizing a research year on the project *Game Theory in the Behavioral Sciences.*[92] I had a lot of exposure to experiments and theory and met many people, like Werner Güth. I started to really think about establishing a lab and to get into laboratory experiments myself. These were key events for me.

Chris Starmer Thank you very much for your contributions. We must stop here. I let things run a little late. But I must say, it has been very interesting to me and enjoyable. We will take a short break now and reconvene at half past.

---

[91] Van Winden visited Caltech from July until September 1986.

[92] Four volumes titled *Game Equilibrium Models* were published in 1991 that cover a variety of topics in economics, biology, sociology, psychology, political science, and behavioral sciences. Most of the contributions were non-experimental, mostly theoretical (traditional non-cooperative game theory). Among the contributors are also researchers who have conducted experimental research such Wulf Albers, *Ron Harstad*, James Walker, Roy Gardner, and *Elinor Ostrom*. **Selten, Reinhard.** 1991. *Game Equilibrium Models*. Berlin; New York: Springer Verlag.

# Chapter 3
# The Growth of a Community

## The German Experimental Community

| | |
|---|---|
| Chris Starmer | In the last session, we were talking very much about early days and the origins of experimental economics. I would like to wind the clock just a little bit further forward and think about what was happening in the late 1970s from then and into the '80s where it seems like various groups were emerging both in the U.S. and particularly in Europe and Germany and holding meetings to discuss experimental economics. Reinhard could you perhaps tell us first a bit about the German experimental society, how it came about, and the meetings that were associated with that. |
| Reinhard Selten | First, I want to go back to the '60s because from the time from '59 to '70 or so, there was a group of people in Frankfurt who continuously worked on experiments. We had always three or four persons working there in changing composition. We got support by the *German National Science Foundation*. *Sauermann* got this [grant], and we had to write reports and so on. I think this was the first continuously working research group for about ten years. At first, we sent everything out as discussion papers only. Then in '67, we published one book about that collected [the papers], mostly in German. Then I think '70, we published a second volume where also a paper by Jim Friedman is in. Then in '72, I think, or it was '71, |



A. Svorenčík, H. Maas (eds.), *The Making of Experimental Economics*,
DOI 10.1007/978-3-319-20952-4_3

we had the big conference in Kronberg near Frankfurt. *Sauermann* organized such a conference.[1] For that purpose, he used—I have to tell the story. He had in the '20s somebody who had left some money as an inheritance to the Department of Economics. They should use this [fund] for research in business administration. But nothing was done with this money. After the war, still was something left, not that much, but because we had war and money reform. But *Sauermann* got the task to administrate this money. Later he was reprimanded because he speculated with this money and tripled it. [some laughter]

Betsy Hoffman: He was reprimanded for that?

Reinhard Selten: Yes. [He] was reprimanded. It was against the rule. It was foundation money, and you should spend the income from that every year. But then they still left him [to administer] this money, and he then used it to finance his first conference. It was an international conference on experimental economics.

Chris Starmer: And when is this?

Reinhard Selten: I think it must have been '71. I'm not –

Charlie Plott: Probably '72.

Reinhard Selten: '72?

Charlie Plott: I think so.

Reinhard Selten: So '72, you [Charlie Plott] were there. [Plott: Yeah] In Germany, I don't know when *Sauermann* founded this society for experimental economics. It was at first only a very small society. He wanted to keep it small because it was founded in order to have somebody who would be responsible for this money, which he had. And so the money was given to this society. He didn't even want to increase the membership.[2]

Chris Starmer: But Charlie, you seemed to know about this quite early on?

Charlie Plott: Which one?

[1] The conference took place in September 1971 in Kronberg, Germany. The contributions appeared as the third volume of the Contributions to Experimental Economics. Foreign participants included Austin Hoggatt, Jim Friedman, and Martin Shubik. The previous two volumes (1967, 1970) had only German authors with a single exception of Jim Friedman in volume 2. **Sauermann, Heinz.** 1972. *Contributions to Experimental Economics. Vol. 3 Beiträge zur Experimentellen Wirtschaftsforschung. Band. 3.* Tübingen: Mohr.

[2] The founding members in 1977 included Otwin Becker, Rudolf Richter, Heinz Sauermann, Reinhard Selten, Reinhard Tietz, Horst Todt, Ulrike Vidmajer und Hans Jürgen Weber. The society opened up only at the beginning of the 1990s.

| | |
|---|---|
| Chris Starmer | This experimental society and the meetings. You knew about this because you were there. |
| Charlie Plott | Well, did I know about them as an organization? No. But we did attend this particular meeting, of course, and got to know them then.[3] |
| Reinhard Selten | It was just an international conference. It was not completely connected to this society. And the society was only much later. I think, in the beginning of the '80s it was generalized. We had increased the membership and it became a greater society. I don't know exactly when this happened, when we opened it up. |
| Chris Starmer | But there were regular meetings of this society? |
| Reinhard Selten | No. There were regular conferences, which were organized by this society. We had about four or five conferences of this kind. I mean, I don't remember. We had five conferences. There were six volumes of papers that were published one for each conference. Once we had two, then there was one on bargaining behavior and coalition behavior. |
| Chris Starmer | And was there very much international participation? |
| Reinhard Selten | Yes. There was a lot of international participation, but also interdisciplinary participation because we also had psychologists all the time in these meetings. For example, one of the participants in this important first, big conference in Kronberg was *Austin Hoggatt*. He should be also mentioned because when I was in Berkeley in '67/'68 I did experiments with him. This was maybe the earliest computerized laboratory. It was an amazing thing. But maybe you don't want to talk about that. |
| Chris Starmer | Well, that is something that I would like to come back to this session to talk about labs. And perhaps we can come back to the Hoggatt lab when we talk about that. I would also like to hear something about the Tucson meetings—did you [addressing Charlie Plott] want to comment on this point? |

[3] Plott attended the 1977 meeting, not the 1972, at least according to the published attendance list. The third International Conference on Experimental Economics took place on August 28th to September 2nd, 1977, in Winzenhohl near Aschaffenburg, Germany. The contributions appeared as volumes seven and eight of the Contributions to Experimental Economics series **Sauermann, Heinz** ed. 1978a. *Bargaining Behavior*. Tübingen: Mohr, ____ ed. 1978b. *Coalition Forming Behavior*. Tübingen: Mohr.

## Caltech, Public Choice, and the "Experimental Bug"

| | |
|---|---|
| Charlie Plott | Well, at this period, there was quite a bit of stuff going on at [Caltech] in 1970, 1971, '72, '73 at the time. I had a large NSF grant. As far as I know, it was probably the only NSF grant for studying experiments [at that time].[4] |
| Jim Friedman | No, it wasn't, Charlie. |
| Charlie Plott | In '71/'72? |
| Jim Friedman | Yes, it was—sorry, wait, no, [inaudible: Plott: yours or theirs was 69-72] it wasn't in '71 or '72. [Plott: yeah] I know somebody who had a grant to do experiments that commenced in '69 and ran for five years. |
| Charlie Plott | Yeah, I don't know who that was. |
| Jim Friedman | Me and Austin Hoggatt.[5] |
| Charlie Plott | Yeah, that was you. [group laughter] Ok, that is right. We can clarify that. Mine was in political economy. Theirs was in duopoly or basically duopoly. [Smith: yeah; Friedman: that's right][XXIV] And I was quite aware of their research. As a matter of fact, it was because I was aware of their research I stayed away from duopoly. I could see both the problems they were having, the complexity of it, and so I said, "That is great, let those guys do that. We are going to study larger groups." We had an NSF grant that would allow us to do a variety of experiments.[6] Betsy and then a whole series of graduate students were doing them. We also had meetings with *Bill Riker* at Caltech at that time. Soon after that started a program called the *Fairchild Fellowship Program* where we could bring people in, of which Vernon was one, John Ledyard |

[4] The first three NSF grants by Charlie Plott were *Political Economic Decision Processes* (period: 11/72-11/74, amount $59,800); *Experimental Examination of Group Decision Processes* with M. P. Fiorina (period: 4/75-4/76, amount: $88,100); *A Laboratory Experimental Investigation of Institutional Influences on Political Economic Processes* (period: 9/78-9/79, amount $95,083).

[5] That was the NSF grant: *Theoretical Research and Collaborative Experimental Research on Micro-economic Games* (period: 2/69–2/71, amount: $101,200) and the NSF grant: *Theoretical Research in Game Theory, Oligopoly, and Individual Behavior* (period: 9/71–9/73, amount: $50,000).

At that time there were two other NSF grants that were used for experimental research. Martin Shubik and Gerrit Wolf's grant was titled *Experimental Economic Psychological Modeling in an Automated Laboratory* (amount $62,300, periods: 12/69–12/70 & 2/72–2/73, total amount: $110,400). Kagel with Battalio and their advisor Basmann had an NSF grant titled *Interpretation Systems for Empirical Economic Theories* (periods 1/72–2/75 & 9/75–9/77, amount: $147,700 & $136,000 respectively).

[6] With *Morris Fiorina*.

was one. We had a whole series of theorists who came through.[7]

And the nice thing about theorists coming through Caltech where we were doing experiments is that they get infected with the experimental bug. They would take it back home with them. I think that that interaction at a fairly high theoretical level with the experimentalists was quite important in terms of placing students later and carrying on a conversation. I would have to get the dates of them, but I know that we had a conference in again '71, maybe late '70. There was probably one in '73. Basically, political science and public economist type things, some game theorists. *Lloyd Shapley* was through.[XXV]

So we had this active group that was taking place from say 1971[8] all the way through the '70s. I knew nothing about the German experiments until I went to this conference. And Reinhard actually did a good job of outlining what they had been doing and what they had studied, and I later used that.

Chris Starme — How did you find out about the conference?

Charlie Plott — They invited me. By that point, we were pretty visible. We had been publishing, doing experiments—certainly in political economy and had agenda stuff that was coming out in the AER, *American Political Science Review*.[9] There was quite a large number of research [projects] that had taken place by that time.[10]

Chris Starmer — John.

[7] *Sherman Fairchild Distinguished* Scholars program started in 1973—incomplete list for the 1970s and 1980s: Vernon *Smith* 1973-74, Bill Riker 1973-74, Melvin J. Hinich 1975-76, John *Ledyard* 1977-78, William A. Brock 1978, David Cass 1978-1979, Richard A. Easterlin 1980-81, Howard Rosenthal 1982-83, Norman James Schofield 1983-84, and Leonid Hurwicz 1984-85.

[8] Plott moved to Caltech in June 1971.

[9] **Cohen, L. and Charles R. Plott.** 1978. "Communication and Agenda Influence: The Chocolate Pizza Design," H. Sauermann, *Coalition Forming Behavior*. Tübingen: Mohr, 329–57, **Fiorina, Morris P. and Charles R. Plott.** 1978. "Committee Decisions under Majority Rule: An Experimental Study." *The American Political Science Review*, 72(2), 575–98, **Plott, Charles R. and Michael E. Levine.** 1978. "A Model of Agenda Influence on Committee Decisions." *The American Economic Review*, 68(1), 146–60.

[10] **Editor:** Here you are clearly referring to the 1977 and not the 1972 conference in Germany. **Charlie Plott**: "By 1972 we were circulating papers. The working paper mechanism was functioning well. Things circulated for years before finally finding print. However, I was not part of the 1972 meeting. Looking over the substance, in retrospect, I can see why there was little or no intersection."

| | |
|---|---|
| John Ledyard | I want to point out that Charlie was also involved in bringing in the *Public Choice Society*.[11] At this point it was serving [Plott: that is true] once a year as an intellectual place for the combination of experimentalists, political scientists, a few weird theorists and game theorists. That was a place a lot of you guys met. |
| Charlie Plott | Actually, I was a president of the *Public Choice Society*,[12] and we used the *Public Choice Society* as a vehicle for bringing the experimentalists together. [Ledyard: you [directing at Smith] came by] That was an organized place for people to meet and share their ideas every year. In fact, the experimentalists at the *Public Choice Society* met together for a long time. [Vernon: yeah] Finally, the *Economic Science Association* split it off and started doing its own thing.[13] |

## The Tucson Meetings, the NSF, and Dan Newlon

| | |
|---|---|
| Chris Starmer | And when did the Tucson meetings start?[XVI] |
| Charlie Plott | Oh, probably in the '80s. |
| John Ledyard | Earlier. |
| Betsy Hoffman | Much earlier. |
| Charlie Holt | Sometime in the '70s. |
| Betsy Hoffman | I think I went to the first one when I was still a graduate student in probably '76 or '77.[14] |

[11] The statement of Purpose of the *Public Choice Society*: "The Public Choice Society was established in 1965 by Gordon Tullock and James Buchanan. Its goal is to facilitate the exchange of work, and ideas, at the intersection between economics, political science, and sociology. It started when scholars from all three of these groups became interested in the application of essentially economic methods to problems normally dealt with by political theorists. It has retained strong traces of economic methodology, but new and fruitful techniques have been developed that are not clearly identified with any self-contained discipline." http://www.pubchoicesoc.org/about_pubchoice.php [Accessed on August 16, 2011].

[12] Plott was the Public Choice Society president from 1976 to 1978. Other PCS presidents with experimental track record are *Elinor Ostrom* (1982-84), John Ledyard (1980-82), Vernon Smith (1988-1990), and *John Ferejohn* (1990-92).

[13] In the 1990s and early 2000s the *Public Choice Society* and *Economic Science Association* held joint annual meetings. The last one took place in Baltimore, Maryland, March 11–14, 2004.

[14] The first workshop took place at the Westward Look Resort, Tucson, March 18-20, 1977. The second workshop, also titled NSF Experimental Economics conference, took place at the Arizona Inn in Tucson, October 19–21, 1979. The third Experimental Economics Workshop took place at the Westward Look Resort, March 27–9, 1984. The Fourth Experimental Economics workshop that took place at the Westward Look Resort, Tucson, Arizona, from February 27 to March 1, 1986.

| | |
|---|---|
| John Ledyard | That was different than the Public Choice Society meetings. |
| Chris Starmer | Yes. |
| Betsy Hoffman | Right. The Tucson meetings, this is pre-ESA. [Smith: Yes] I started going to the *Public Choice Society's* [meetings] when I was a graduate student as well. I started going to the Tucson meetings as well. I can remember one in particular where everyone working in this area could sit around the boardroom at the Arizona Inn.[15] I remember the first Tucson meeting I went to [Smith makes a confirmatory sound], we were all around. There were probably at most 30 of us, maybe 25. We all sat around one big table at the Arizona Inn, and to me, that was a defining moment in the formation of organized experimental economists.[16] |
| Chris Starmer | And who were the organizers of that? |
| Betsy Hoffman | I believe Vernon organized that. |
| Vernon Smith | Yes. I got funding, NSF funding for a seminar in experimental economics, workshop in experimental economics in 1978, again in '79. |
| Betsy Hoffman | So maybe it was '78. |
| Vernon Smith | Yes. And I don't know which—one was at the Arizona Inn and one was at the Westward Look.[XVII] |
| Betsy Hoffman | I think I went to both of them. |
| Vernon Smith | Yes. And there was an addition. *Martin Shubik* was maybe at the first one. I'm not sure if he was at the second one or not. Charlie [Plott], *Fiorina, Noll* – |
| Betsy Hoffman | *Richard Thaler* was at the one at the Arizona Inn.[17] |
| Vernon Smith | Yes. *Mark Isaac*. [Hoffman: Right] But the first funding I got was in 1962 from the NSF, and I renewed that.[18] Howard Hines was the head of the division of social science at NSF and it must have been about 1965, 1966. I was regularly going to Washington. There was a lot of NSF funding of graduate |

[15] Only the second meeting in 1979 took place at the Arizona Inn.

[16] The first workshop had 24 invited participants (three could not come). The second workshop had 34 participants (with five from Arizona, nine from Caltech) and the third from 1984 had 54 listed participants.

[17] The 34 participants in 1979 were (Witness Seminar attendees are in italics): Battalio (Texas A&M), Burns (Australia), Dawes (Oregon), Easley (Cornell), Ferejohn (Caltech), Green (Washington), Grether (Caltech), Groves (UCSD), Harstad (Illinois), *Hoffman* (NW), Isaac (Caltech), *Kagel* (Texas A&M), *Ledyard* (Northwestern), Marrese (Northwestern), Murningham (Illinois), Nelson (Caltech), Newlon (NSF), Noll (Caltech), Palfrey (Caltech), Philips (Cornel), *Plott* (Caltech), Rachlin (STONY Brook), *Roth* (Illinois), Rotchschild (Wisconsin), Stadon (Duke), Thaler (Cornell), Wilde (Caltech), Williams (Indiana). From University of Arizona came – Alger, Auster, Cox, *Smith*, Taylor, and Walker.

[18] NSF grant G-24199 titled *Behavior in Competitive Markets* with a total award of $39,500 and duration 7/62-7/64.

students at that time and I was involved.[XVIII] I remember Howard Hines at one of those told me that the only reason why they have been able to fund me was by just simply ignoring referee reports that dismissed experiments. Then, apparently, they had a flexibility they don't have now.

Betsy Hoffman They still have it.

John Ledyard They don't use it.

Betsy Hoffman They don't exercise it. *Dan Newlon* exercised it quite a bit.

Vernon Smith Yeah. So if the referee reports took it seriously, then they took seriously the referee reports. [laughs]

Chris Starmer Can I just reserve this to a little bit later for the discussion of the funding. I think that is a very interesting topic, but I would like to focus on that a little bit later. Can I just bring you back to the Tucson meetings, and Betsy described it as a defining moment. I wonder whether you can say more about that or whether other people had a similar view of that.

Charlie Plott I considered it a defining moment in one sense, but remember that meetings with the *Public Choice Society*, this was an organized group, and they were meeting annually. We continued to meet even after ESA was started. Basically, you are watching two parallel things. The experimentalists are growing, and then they decided to actually organize themselves into a society. That was controversial, by the way, because we were afraid we would get ghettoized. If you had a group and a journal called *Experimental Economics*, would all of the journals then shove us off into that, so there was some controversy about whether that was a good idea to start a separate journal or not.[XXIX]

And so obviously, the decision was to create a society and then soon after that, start a journal. Then after that, you start seeing this more organized thing. This must have been—what? In '80, '81?

Betsy Hoffman It was mid '80s. It was actually a little bit later than that.

Charlie Plott In '81 or '82, something like that?

Betsy Hoffman I think I was already at Wyoming when ESA started, which would have been '85/'86.[19]

Vernon Smith It was '84, '85, I think.

[19] Hoffman was a Professor of Economics at University of Wyoming from 1986 to 1988.

## Bringing Experiments to a Larger Audience

| | |
|---|---|
| Frans v. Winden | I have a question related to this. Apparently, the *Public Choice Society* was a vehicle to bring experimental economics to a larger public. Did people actually try to bring it into the *Econometric Society* meetings? |
| Betsy Hoffman | Yes. [Plott makes a confirmatory sound] In fact, I met Reinhard at the 1980 International – |
| Frans v. Winden | But there were no real [experimental] sessions, I think. |
| Betsy Hoffman | I'm not sure. |
| Frans v. Winden | I know from the *Public Choice Society* [that there were experimental sessions], but I'm not aware that the *Econometric Society* meetings showed real [experimental] sessions. |
| John Ledyard | Not for a long time. |
| Betsy Hoffman | But I did present—you [addressing Selten] were my discussant. I presented a paper from, I can't remember which one it was, but I think it was from the long line of papers with either *Ed Packel* or *Matt Spitzer*. [Selten: Yes] And Reinhard was my discussant at the 1980 International Econometric Society in Aix-en-Provence.[20] |
| John Ledyard | World Congress. |
| Charlie Plott | We gave a paper on the airport experiments and the airport slot allocation; it was around 1979 or so at the AEA.[21] We were able to get on the [program] of the larger societies' meetings. |

[20] The fourth *World Congress of the Econometric Society* took place from August 28 to September 2, 1980. John Ledyard was one of 49 members of its Program Committee. The session titled *Labor Supply* had a presentation by Orley Ashenfelter on "Discrete Choice in Labour Supply: The Determinants of Participation in the Seattle and Denver Income Maintenance Experiment." Plott presented in the session on *Economics of Information* paper titled "The Behavior of Markets with Insiders: A Laboratory Experimental Examination of Rational Expectations Models," which was done jointly with Shyam Sunder, Chicago. There was one Session on Experimental Economics with the following two papers "An Experimental Test of Several Solution Theories for Cooperative Normal Form Games," by Richard D. McKelvey and Peter C. Ordeshook, California Institute of Technology, and "An Experimental Study of the Effect of Exogenous Voting Costs on the Decisions of Majority Rule Committees," by Elizabeth Hoffman, Northwestern University and Edward W. Packel, California Institute of Technology and Lake Forest College. The chairman of this experimental session was Reinhard Selten, University of Bielefeld, Germany. A field experiment "Complete Demand Systems for India: A Series of Experiments with Linear Expenditure System, under Alternative Specifications" was presented by G. V. S. N. Murty of the Sardar Patel Institute of Economic and Social Research, Ahmedabad, India.

1981. "Program of the Econometric Society World Congress." *Econometrica*, 49(1), 247–76.

[21] Plott presented the paper "Allocation by Committee: The Distribution of Airport Capacity" at the 1980 AEA meetings in Denver. It was presented in a session on Models of Antitrust and it was the only experimental paper in that session.

| | |
|---|---|
| Frans v. Winden | Reinhard, I met you in '76 at the *Econometric Society European Meeting* in Helsinki for the first time. Do you remember if there were any experiments being presented? I'm not aware of any.[22] |
| Reinhard Selten | I don't remember what I presented but there were some things discussed by us. In the discussions, it appeared. But... |
| Frans v. Winden | Because I think it is interesting if a society like the *Econometric Society* didn't pick that up. |
| Jim Friedman | I think experiments, I was just checking here... |
| Betsy Hoffman | It did. |
| Charlie Plott | It did. |
| Jim Friedman | And in 1975, *American Economic Association* meetings, *Hoggatt* and I had a paper in a session. I cannot remember if it was in total an experimental session because I don't remember the surrounding circumstances. But my guess would be it was.[23] |
| Charlie Plott | My impression is that we were sprinkled through those meetings. [Smith: ehm] *Econometric Society*, I was going to those meetings regularly, and at that time, if I gave a paper, it would have been experimental, and I was going to those meetings regularly. So I would have to go back and see.[24] |

[22] **Editor**: The Program from Helsinki does not list Selten or van Winden. See: 1977. "Program of the Econometric Society Summer European Meeting, August 24–27, 1976, Helsinki, Finland." *Econometrica*, 45(1), 252–56. **Frans van Winden**: "However, Econometrica didn't show the Contributed Papers of the meeting; mine was a contributed paper, entitled "The Interaction between State and Business: A First Approach" (Report 76.09, The Economic Institute of Leyden University)."

[23] The meeting in Dallas, Texas, December 28–30, 1975, had a session titled *Experimental Economics*. Three papers were presented—the one mentioned by Jim Friedman jointly with Austin Hoggatt on Pricing Signaling in Experimental Oligopoly; one paper by Vernon Smith titled Experimental Economics: Some Theories and Results; and a paper by Roger Noll On the Pricing of Public Goods: Public Television Programming. The Chairman was Martin Shubik and commentators Charlie Plott and Michael Rothschild.

**Ferejohn, John A. and Roger G. Noll.** 1976. "An Experimental Market for Public Goods: The Pbs Station Program Cooperative." *The American Economic Review*, 66(2), 267–73. **Hoggatt, Austin C.; James W. Friedman and Shlomo Gill.** Ibid. "Price Signaling in Experimental Oligopoly." 261–66, **Smith, Vernon L.** 1976. "Experimental Economics: Induced Value Theory." *The American Economic Review. Papers and Proceedings of the Eighty-eighth Annual Meeting of the American Economic Association*, 66(2), 274–79.

[24] In the period until 1978 Plott presented at the Econometric Society in 1966, 1969, 1970 and was a discussant 1967, 1968, 1972; 1969. 1973, 1975 as discussant and session chairman. **Plott, Charles R.** "Plott Papers," *personal archive of Charles Plott*. California Institute of Technology, For instance, the Program of the 1979 North American Summer Meeting of the *Econometric Society*, Montreal, Canada, June 27–30, 1979 session Experimental Studies of Uncertainty and Information had the following presentations: "*Risk Aversion and Portfolio Selection: Experimental Evidence*," Ronald M. Harstad and Edward M. Rice, University of Illinois (Champaign-Urbana). "*Professional Diagnosis vs. Self-Diagnosis: An Experimental Analysis of Markets with Uncertainty*," Charles R. Plott and Louis Wilde, California Institute of Technology. "*Game Theoretic Models and the Role of Information in Bargaining*," Alvin E. Roth and Michael

| | |
|---|---|
| Chris Starmer | Vernon. |
| Vernon Smith | Well, my 1976 paper on the theory of induced valuation was given at an AEA meeting—and [it was a] good thing. It probably would have never gotten published otherwise because it hadn't been refereed.[25] [some laughter][XXX] |
| Betsy Hoffman | Was that the annual meeting? |
| Vernon Smith | Yeah. |
| Jim Friedman | *Papers and Proceedings*, yeah. |
| Vernon Smith | It was published as a short piece in the *Papers and Proceedings*. |
| Charlie Plott | I was the discussant of that actually. |
| Vernon Smith | You were there. And *Martin Shubik* was engaged. |
| Charlie Plott | *Debreu* was in that audience, too. |
| Vernon Smith | And *McFadden. Dan McFadden.*[XXXI] |
| Charlie Plott | Maybe |
| Chris Starmer | Betsy. |
| Betsy Hoffman | I was going to say I was regularly going to the *Public Choice Society* meetings and the *Econometric Society* meetings and the *American Economic Association* meetings. There were sprinklings of a lot of papers at the *Public Choice Society* when Charlie was president and then when *Lynn Ostrom* was president, she carried on too, as well.[26] In fact, she probably was another person who brought my career along by getting me to organized sessions [Plott: That is true] and to be on the organizing group of the *Public Choice Society*. So there were a sprinkling of papers at that Econometric Society, lots of papers at the *Public Choice Society*, and a sprinkling of papers at the *American Economic Association* meetings. |

W. Malouf, University of Illinois (Champaign-Urbana). "*Consumer Demand Behavior with Pigeons as Subjects: Limits to Rationality*," John H. Kagel, R. C. Battalio, L. Green, and H. Rachlin, Texas A and M University. The discussants were: Phillip Dybvig, Yale University; Robert G. Wolf, Boston University; Ehud Kalai, Northwestern University; Steven A. Mathews, University of Illinois (Champaign-Urbana).

[25] **Smith, Vernon L.** 1976. "Experimental Economics: Induced Value Theory." *The American Economic Review. Papers and Proceedings of the Eighty-eighth Annual Meeting of the American Economic Association*, 66(2), 274–79. It appeared earlier as a working paper ____. 1975. "Experimental Economics: Theory and Results." *Social Science Working Paper, California Institute of Technology*, 73(January), 1–16. which was based on ____. 1973. "Notes on Some Literature in Experimental Economics." *Social Science Working Paper, California Institute of Technology*, 21, 1–27.

[26] Four volumes titled *Game Equilibrium Models* were published in 1991 that cover a variety of topics in economics, biology, sociology, psychology, political science, and behavioral sciences. Most of the contributions were non-experimental, mostly theoretical (traditional non-cooperative game theory). Among the contributors are also researchers who have conducted experimental research such Wulf Albers, *Ron Harstad*, James Walker, Roy Gardner, and *Elinor Ostrom*. **Selten, Reinhard.** 1991. *Game Equilibrium Models*. Berlin; New York: Springer Verlag.

John Ledyard — I was chair of the [North American] *Econometrics Society* program committee in 1981.[27] And to be perfectly honest, there weren't that many experimental papers being submitted for inclusion in the program. There were certainly not enough to create a separate program for them. Whatever came in, and there wasn't very many, got allocated to the way you would want them allocated [referring to Plott], which is to the area in which they were involved in and included, if they were good enough. This was a speck in the ocean at *Econometric Society*. It was not a big deal at that point.

Betsy Hoffman — I think I may have given one of my Coase papers at that meeting.[28]

John Ledyard — Yeah, could be.

## Struggle for Acceptance or Standard Battles with Referees?

Chris Starmer — So I get the impression that there is a tension between on the one hand simply wanting to think of experimental economics as a tool, which you apply in different areas, and not think of it as in some sense some specialized pursuit, which might make it ghettoized. And on the other hand, experimental research is wanting to have an identity through some society, some recognition that they are in a group perhaps with critical mass. [Hoffman: Yes] [Does] anyone have any reaction to that way of putting things?

Charlie Plott — The tension?

Chris Starmer — Yes.

Charlie Plott — No, I don't recall any particular tension. The only tension that I saw was the possibility that if you started a journal, then you would be kicked out of the major journals and told to go get specialized. I don't recall any identity crisis or any kind of worry [along those lines]. I think that all of us were under fire from the specialized referees that all of us got to know well. But other than that, it is just a standard battle. I must say that I would continually during this period, hear comments about how controversial experimental economics was. I gave hundreds of lectures and seminars. I have never, ever had somebody out in the audience being critical of experimental economics—ever. And so you hear about these criticisms, but whatever was out there was not surfacing in a room where I had the mike.

---

[27] Plott moved to Caltech in June 1971.

[28] The paper was presented at the 1980 World Econometric Meeting.

Jim Friedman — Charlie, I have a feeling that what you are addressing expresses the notion: I think what you are doing is good and interesting, but I think most people won't agree, and most people think that way. So the de facto situation was that there was a high degree of acceptance and a kind of skepticism about whether that acceptance was broadly based when, in fact, it was broadly based. [Smith: ehm]

Charlie Plott — So that was the tension that people were reporting. I just never saw it.

Jim Friedman — No. I think the sense of tension came from a lot of people thinking that experimentation wouldn't be well received and just simply being wrong about that.[XXXII]

Charlie Plott — Yeah. That is interesting.

Betsy Hoffman — Well, I beg to disagree because it was really coming out in the referee's reports. Yes, people were very polite, and I gave hundreds of seminars, too. And people were very polite and asked for copies of my papers, but it surfaced in the referees' reports. I have a paper that never got published and which actually people have started asking me for copies of it. And I had to go find a copy of it. This paper was doing an experimental test of the Groves-Ledyard mechanism. My husband, who doesn't have as many papers as I do, feels very strongly about the fact that the paper never got published.[29]

We just kept writing it and revising it and doing more experiments and sending it out to be refereed. Every time we got back a referee report, we were told that it didn't have a sufficient theoretical basis. Well, our theoretical basis was the Groves-Ledyard paper.[30] But that was not—we just kept getting told over and over and over again that simply running experiments to show that a particular theoretical model would lead to a tâtonnement mechanism would lead to the predicted result was not sufficient. We never finished the paper. I became an administrator and basically got tired of sending off a paper and continuing to have it rejected.[31] Just a year

[29] Actually, there is a later version: Binger, B., Hoffman, E., and Williams, A., (1987) "Experiments on a Tatonnement Mechanism for Allocating Public Goods," joint meeting of Public Choice Society and ESA., March 1987. Binger, B. and Hoffman, E. (1983) *Non-linear Prices and the Optimal Allocation of Public Goods*. Draft prepared for presentation at the Public Choice Society Meetings, March 24–26, 1983, Savannah, Georgia.

[30] **Groves, Theodore and John O. Ledyard.** 1977. "Optimal Allocation of Public Goods: A Solution to the" Free Rider "Problem." *Econometrica*, 45(4), 783–809.

[31] That was in 1989 when Hoffman became the Associate Dean and the MBA Program Director at the Karl Eller Graduate School of Management, University of Arizona.

ago, I started getting requests for this paper, and I had to go find the most recent version.

Now, you have got about 10 different versions, and this was long before we had everything on our computers. I had to literally go in my boxes in my basement and find the most recent version of this paper, which has now been scanned is available in PDF form for anybody who wants it. But I think that is an example of the kind of tyranny, as it was that was exacted on some of us through the referee process.

Chris Starmer Vernon.

Vernon Smith Well, Betsy, I think also you were developing in that paper a process for implementing the Groves-Ledyard, and that is the thing that was not well modeled and is still not well modeled in economics.

Betsy Hoffman No. That is true.

Vernon Smith I mean, the means by which groups using rules to discover things,[XXXIII]

Betsy Hoffman But the idea that you could do these things

Vernon Smith those models—look how long it took for the double auction to get modeled, and even then it has a lot of limitations. All kinds of really good people tried and worked on it. *Bob Wilson*, for example, was one of the first people to make progress.[32]

Chris Starmer John [Kagel], then Charlie [Holt].

## Educating Editors

John Kagel Well, we were trying to publish papers with animals as subjects in economics, and so we got some, at times, very nasty referee reports. We took it on ourselves to educate editors. So we would write a letter back and just try to outline what we were trying to do. Not that it would change any decisions, but that these people needed education.

[32] **Wilson, Robert.** 1985. "Incentive Efficiency of Double Auctions." *Econometrica*, 53(5), 1985. **Editor**: Did you have other Wilson's papers in mind? **Vernon Smith**: "This is one of them; but there was also one in a collection in honor of Arrow. I think this latter one was the one in which he showed that the complete information game-theoretic model of the DA led to failure in the end game of a DA trading period. He saw this as an inescapable consequence of complete information—a contradiction—and I always felt this was dead right, that it converged in experiments because the final traders were uncertain about the circumstances and even the number of people left who could feasibly trade." See also **Smith, Vernon L.** 2008c. *Rationality in Economics: Constructivist and Ecological Forms*. Cambridge: Cambridge University Press.

Chris Starmer: So there was a feeling that referees were in some sense misunderstanding what people were trying to do?

John Kagel: Yeah. I remember distinctly, we sent a paper to *Science*. We were very excited. We were doing this stuff with animals and were getting consumer choice optimization as economists defined it, etc. And I don't have the referee reports [anymore], but they were like you can't do this. You are crossing a line that shouldn't be crossed. It was almost a religious response by some sort of a fundamentalist. What do you do? Anyway, we found sympathetic editors, and *Bob Clower*[XXXIV] was a particularly sympathetic editor at what used to be called the *Western Economic Journal*, and we got some of our papers accepted there before they [subsequent papers] went into major journals.[33]

Chris Starmer: Charlie [Holt].

Charlie Holt: I remember when I put together a bibliography of experimental economics in the year 2000. That was the idea of the *Y2K bibliography*.[34] I tracked the percentage of papers that were published in the AER, and it was pretty high. [Smith: ehm] Experimentalists had good luck in the AER in the '80s and even more so in the '90s. Of course, there was a big increase in the mid '90s after Reinhard's prize.[35] That was a big plus. [Hoffman: Yeah, that helped][XXXV] But I think we did pretty well. When I think back about some of my experience, it is the kind of work that is understandable to people in other areas. I think NSF funded economics experiments to a large extent because they could present it as science to other parts of NSF.

I remember at Minnesota[36] applying for a grant within the university that would be—against people in other

---

[33] Kagel and Battalio published two experimental papers in the *Western Economic Journal* in the 1970s. The first one reported on their token economy experiments and received the Best article Award for 1973. The latter was their first published article with animal experiments.

**Battalio, Raymond C.; John H. Kagel; Robin C. Winkler; Edwin B. Fisher; Robert L. Basmann and Leonard Krasner.** 1973. "A Test of Consumer Demand Theory Using Observations of Individual Consumer Purchases." *Economic Inquiry*, 11(4), 411–28, **Kagel, John H.; Raymond C. Battalio; Howard Rachlin; Leonard Green; Robert L. Basmann and W. R. Klemm.** 1975. "Experimental Studies of Consumer Demand Behavior Using Laboratory Animals." Ibid.13(1), 22–38.

[34] The Y2K Bibliography of Experimental Economics and Social Science is available online at http://people.virginia.edu/~cah2k/y2k.htm [Accessed on March 31, 2015]. This bibliography lists over 2000 publications, plus about 500 discussion papers in experimental economics and social science (updated December 29, 1999).

[35] Nobel Prize of 1994 together with John Nash and John Harsanyi.

[36] Charlie Holt spent the period 1976–1983 at University of Minnesota.

disciplines. And *Leo Hurwicz* came back and told me that the people on that panel were really excited because they could sense what the issue was, what the question was, and what the procedure was going to be. It was something you could explain.

Chris Starmer Charlie [Plott].

Charlie Plott On the referees, this editor education process is something we were doing constantly. I think that I never had a paper rejected that I didn't carry on a conversation with the editor about. And the editors were fantastic, by the way. They would listen. In *Econometrica*, for example, they would take the referee report and sent both referee reports together with your objection to one of the other referees, and they would very frequently go your way.[37] I found the editors really, really receptive in the major journals. We had success in that way. Now, as time went on though, you began to get rejections from experimentalists who themselves had philosophical preconceptions about how you do experiments.

And so then you would not get rejections from a knee jerk person who really didn't understand what you were doing. You were getting an objection from somebody who has a philosophical objection with what you are doing, and actually knows something about it. During this period, you start to find the character of the referee reports changing. And dealing with the experimentalists, they are a little bit more intractable because it is almost a religious problem that we are beginning to see at this point. There is this evolution that takes [place]—but I think that the idea that there was non-acceptance is not exactly right. And I think Jim might be right.

The skepticism is about how receptive it might be. But if you actually look at what is happening in the journals, the journals are quite receptive.

Al Roth My experience coincides with that. When my early papers were being refereed by game theorists, I think they were easily received. As the referring became more professional and as referees' reports were written by experimenters, sometimes I would have doctrinal disputes.

[37] Econometrica had the following editors until 1992: Ragnar Frisch (1933-1954), Robert H. Strotz (1955-1968), Franklin M. Fisher (1968-1977), Hugo Sonnenschein (1977-1984), Angus Deaton (1984-1988), Andreu Mas-Colell (1988-1992).

## Amsterdam: Bridging Experimental Cultures

Chris Starmer: I would like to just change slightly back to a theme we were on a little while ago, which was about the evolution of groups and workshops. And I would like to ask Frans about the development of the workshops in Amsterdam and how that came about.

Frans v. Winden: We started with a series of workshops in the beginning of the '90s and had the first one in '92. We had six of them in a row. It was instigated by a very important new grant by the *Dutch Science Foundation*, the NWO, the Pioneer Grant.[38] And that enabled us to start a research group, CREED.[39] We had also an obligation actually to spread the word so to speak. NWO wanted us to teach other people, particularly in the Netherlands of course, to deal with experimental economics and to run experiments. So we started this series of the *Amsterdam Workshop on Experimental Economics.*[40]

In '93, we held one here actually in this room. I have a picture with me. Al Roth was there, Charlie was there, Reinhard was there, this Charlie [Holt] was there, and John [Kagel] was there, and Vern was not there.[41]

Vernon Smith: I was at one or two of them.

---

[38] The proper English translation of the NWO is *The Netherlands Organization for Scientific Research*. The grant entitled *Laboratory experiments in political economics* lasted from 1991 until 1996.

[39] *Center for Research in Experimental Economics and political Decision-making*.

[40] The *International Journal of Game Theory* (IJGT) published a special issue on the first Amsterdam workshop. **Harstad, Ronald M.** 1996. "Special Issue on Laboratory Investigations of Expectations in Games: The Amsterdam Papers." *International journal of game theory.*, 25(3), n.p.

[41] The full list of 1992 Amsterdam workshop attendees—36 participants (Albers, Antonides, de Beus, Bohm, Bolton, Brandts, van Damme, Forsythe, Güth, Harrison, Harstad, Hessing, Hey, Hoffman, Kagel, Levin, Loomes, McKelvey, Nagel, Noussair, Olson, Ostmann, Palfrey, Potters, van Raay, Robben, Schram, Selten, Smith, Sonnemans, Starmer, Sugden, Tijs, de Vries, Wilke, de Wit) and four guests (Drissen, Mazza, Otterman, and Perdeck).

The 1993 Amsterdam workshop had 43 participants and five guests (e.g. Forsythe, Hey, Camerer, Cox, Holt, Damme, Keser, Fehr, Ledyard, Loomes, Porter, McCabe, Morton, Roth, Olson, Plott, Selten, Starmer, Weber). The 1994 Amsterdam workshop had 49 participants (e.g. van Damme, Albers, Andreoni, Bohm, Frey. Fehr, Bohnet, Grether, Kagel, Keser, Knetsch, Loomes, Selten, Smith, Starmer, Weber). The 1995 Amsterdam workshop had 46 participants (e.g. Albers, Bolton, van Damme, Fehr, Forsythe, Dan Friedman, Gächter, Harrison, Harstad, Hey, Holt, Kagel, Morton, Keser, Nagel, Olson, Selten, Plott, Sugden, Rutström, Sadrieh).

From the participants of the Witness Seminar only Jim Friedman and Steve Rassenti did not participate. The former wasn't active as an experimentalist at that time, the latter primarily specializes in applied contract research in collaboration with Vernon Smith.

Frans v. Winden What was important is that the German society was in practice mainly for Austrian, German, and Swiss people [group laughter]; it was German speaking.[42] I was not aware of it actually that something like that was going on. Apart from the German speaking countries, there were not that many people aware of these things in Europe. We were an odd corner in terms of the experimental economics world. In that context, the Amsterdam workshops in experimental economics played an important role because they brought together all—we had money to invite people.

We brought all the top dogs in experimental economics at that time over the years to Amsterdam. We invited gradually all the people in Europe who were interested. That played an important role to get experimental economics known in Europe on a larger scale and to get people started to do experiments. Then later on, after '97 when we stopped that, I got this grant from the *European Union* to start a network on experimental economics called ENDEAR—*European Network for Experimental Economics and Applications*.

Apart from stimulating joint research projects, we organized a series of summer schools, together with workshops.[43] The summer schools were important to train young people. The first one was in Bari, organized by *John Hey*, and many people are still referring to that one. Actually, we were at a conference in China,[44] Reinhard [Selten] and I, and people there were also referring to that conference because many young people, the non-German people mainly, joined forces there, got into joint projects, and learned to do experiments, and built a network. Many of those people have now established positions, quite a few actually, as professors and are doing experiments.

[42] *Gesellschaft für experimentelle Wirtschaftsforschung* [Society for Experimental Economics Research] founded in 1977.

[43] Van Winden was the project director of the The European Network for the Development of Experimental Economics and its Application to Research on Institutions and Individual Decision Making from 1998-2003. The Network consisted of the following institutions (local coordinators in parentheses)—University of Amsterdam (van Winden), Universitat Pompeu Fabra in Spain (Bosch-Domenech), University of York in the UK (John Hey), University of Bonn (Reinhard Selten), Ludwig Boltzmann Institute for Economic Research in Vienna (Ernst Fehr), Humboldt University in Berlin (Werner Gűth), Instituto de Analisis Economico in Barcelona (Jordi Brandts) and the Centre for Rationality and Interactive Decision Theory at the Hebrew University (Shmuel Zamir). Two main research goal were addressed—1) the development and influence of institutions; 2) the fundamental study of individual decision making in economic situations.

[44] The week before the witness seminar.

The Amsterdam workshops in experimental economics, followed by ENDEAR and later ENABLE,[45] which was a network in behavioral economics, were quite influential in this respect. ENABLE also organized three summer schools.

In a period of ten years, we had something like seven summer schools. And that I think helped quite a bit in bridging the German experimental community with the American experimental economics community.

## Internationalization and the Need for a Journal

Chris Starmer: And what was your relationship to ESA?

Frans v. Winden: I had been at Caltech in '86, and then in '91 Arthur Schram and I got the pioneer grant, and we could start CREED. We made a tour to the United States, and also we went to Bonn to get an idea of how these labs were run. We went to Caltech and Tucson, and to Richmond to meet *Ron Harstad*. We saw Al Roth in Pittsburgh, and also *Shyam Sunder* was there.[XXXVI] I think he was still running his zero intelligence trader experiments at that time. [Hoffman: Still is] These were the major centers in the USA. We took the installation of the [CREED] lab very seriously. Sorry, what was the question again? [laughter]

Chris Starmer: The relationship of the Amsterdam group meetings to ESA.

Frans v. Winden: That was another occasion where we really tried to link up with what was happening in the United States and in Germany. I decided to go to the ESA meetings. Those started in '86. At some point, it was in '94, I think, I became a member of the Executive Board. Thus I got also involved in the decision making process and how people worked it out.[46] Then there was another important event in '97 in Bonn, where

[45] Van Winden was the project director of the Research Network grant ENABLE: The European Network for the Advancement of Behavioural Economics 2004-2008. The Network consisted of following nine institutions (local coordinators in parentheses)—Universiteit van Amsterdam (Frans van Winden), Centre for Economic Policy Research (Ian Jewitt), Universität Zürich (Ernst Fehr), Stockholm School of Economics (Magnus Johannesson), Institut d'Economie Industrielle (Bruno Biais), Universität München (Klaus Schmidt), Universität Mannheim (Martin Weber), Harvard University/MIT (David Laibson), and Princeton University (Roland Bénabou).

[46] Van Winden was a member of the executive board of the *Economic Science Association* from 1994 until 2001.

a conference on *Bounded Rationality* took place.[47] There the decision was made to—actually the decision was made the year before in Tucson by the ESA at the Westward Look—to internationalize ESA. And we had a discussion in Bonn whether people in Europe would be willing to support that.

There was not an immediate acceptance. There were some discussions, but in the end the internationalization was accepted. That meant that there would be an international conference every third year, and local conferences the other two years.

Chris Starmer — Can you say a bit about what the discussions were about? What were the issues as far as people were concerned?

Frans v. Winden — I think that some of the issues discussed also surfaced at the Amsterdam workshops. There was a different conception about how to do experiments and maybe also some fear that because of the greater number of people involved in the United States their views might perhaps dominate too much. I remember from Public Choice conferences that this issue of dominance also turned up when Public Choice emerged as a research field in Europe in the '70s and '80s. I was involved in the start of the *European Public Choice* movement.[48] Also in that case people felt that it was important for Europe to develop the field without being influenced too much by researchers from the United States. That played a little bit of a role in the ESA context too. [Notwithstanding the fact that] there was already the German [experimental] society that had been organizing conferences for many years. So there were questions whether this [internationalization] was the right thing to do. At least that is my perception of these discussions in Bonn. But in the end, people thought it was a good idea. It was decided to go for it. Then the next year, there was also the first issue of *Experimental Economics*.

Chris Starmer — Okay. We will come to *Experimental Economics* in a moment. Charlie, you were waiting to get in.

Charlie Holt — Yeah. I was going to provide a little bit of a backstory related to *Experimental Economics*, but should I come back to that later?

Chris Starmer — Well, no. I would be happy if you wanted to start that now.

---

[47] The workshop *Theories of Bounded Rationality* was organized by Bettina Kuon, Abdolkarim Sadrieh, and Reinhard Selten. It took place in Bonn on May 6–10, 1997.

[48] Van Winden was member of the executive board of the *European Public Choice Society*, 1984-1995, and president of the EPCS in 1986/87. The EPCS was founded in 1972 by a small group of scholars under the leadership of Peter Bernholz of the University of Basel.

Charlie Holt When I was the president of ESA,[49] we brought a couple of proposals in to start a journal. One was very informal. And one was well developed. It was from Kluwer. Specific proposal to start a journal, and twice we talked about it in the Executive Committee meetings, and it got blocked just because of this argument about being ghettoized. We were having pretty good luck in the mainline journals, too, at that time. That was another counter argument but very effective. Then I met *Arthur Schram* at one of these conferences, probably in Amsterdam.[50] And Arthur and I just decided to go ahead and start a journal. And just do it. [Smith: That is how you do experiments, just do it] And then we took it to the ESA and see if they wanted to be connected or not. And so first, we went to Vernon and asked him to be an advisor, and he said yes.[XXXVII]

And about the same time, Arthur and *Zac Rolnik* who was the publisher approached Reinhard, and he said yes, he would be glad to be an advisor. And then we went to Al Roth. He said, "Vernon is a good man, Reinhard is a good man, I will be an advisor." And Charlie Plott, to his credit, bargained a fair amount. He was pushing really hard for colored graphs, and even the price of what each page cost. And the publishers pushed back on that. [some laughter] In the end, we had to give up on that. We went to ESA, and we said we are going to start a journal. But if it can be an ESA journal, that would be better. *Tom Palfrey* who was president at that time really grabbed the ball. He got a Caltech lawyer.[51] Tom said it can be one of our ESA journals.

We are not just going to have one journal, we might have many. He was thinking big. Tom got the Caltech lawyer to also work on this process of internationalizing the ESA. They negotiated everything right down to what the subscription rate on the journal would be like 12 years later based on the CPI change or whatever. Tom was actually a key person in that internationalization process. We had a meeting in Germany, maybe it was in Mannheim, a dinner. [confirmatory sounds of

[49] ESA's presidents in the following periods were: 1986-1987: Vernon Smith 1987-1988: Charles Plott; 1988-1989: Ray Battalio; 1989-1991: Elizabeth Hoffman; 1991-1993: Charles Holt; 1993-1995: Robert Forsythe; 1995-1997: Thomas Palfrey; 1997-1999: James Cox; 1999-2001: Andrew Schotter; 2001-2003: Colin Camerer; 2003-2005: Ernst Fehr; 2005-2007: John Kagel; 2007-2009: James Andreoni; 2009-2011 Tim Cason; 2011-2013 Alvin Roth; 2013-5: Jacob Goeree; 2015-2017: Yan Chen.

[50] Actually they met at the World Econometric Congress in Tokyo "in 1995".

[51] Michael Keller, a lawyer at Caltech's Patent Office.

Selten and van Winden[XXXVIII]] There was a little bit of tension there, because the Germans had their association before. The issue was would they have to pay for this journal. I don't know quite what the issue was.

But anyway, *Zac Rolnik* who was the publisher was just brilliant at coming up with a solution. He recognized, of course, that the publishers don't make much money on selling journals to individuals. He said, "I tell you what. We are going to give free subscriptions to the German association for a couple of years." And that was the end of that discussion.[52] The connection to Arthur was important. And that was also very important in making it an international journal, helping transform ESA into an international association, and connecting up with the Germans, which was where all the tradition was in Europe at that time.

Charlie Plott — Actually, we do have some color in those journals now. You get two pages per issue!

Charlie Holt — Speaking of the Dutch ESA connection, the cover of that journal. The idea was to get something that had sort of Amsterdam [in it.] These bright yellow and white artistic colors connected to some ESA type step function. That was the idea behind the graphic on the cover, which they just recently slightly redid.

Chris Starmer — And there were fears about ghettoization. How do those fears look now? Were they valid fears or not?

## A Separate Journal: A Ghetto or a Premium Site to Publish?

Reinhard Selten — I would have started the experimental journal long before, but I also saw that I wouldn't get the support of you and of others for doing this because in America there was a strong feeling that these papers should get into the ordinary journals. And as you [referring to Plott] said, these experimental papers should not be pushed into specialized journals. I respected this. Otherwise, I would have had the opportunity to found such a journal long before then. At one point in the [early] '90s, I got calls by five publishing houses who all wanted to establish a journal on experimental economics at the same time. I told all

[52] The journal *Experimental Economics* was launched in 1998.

of them that there are five publishing houses trying to do this. And then nobody did it because it would have been a catastrophe, if five such things would happen at the same time. [some laughter] I didn't do this.

Then with Springer we did a section [on experimental economics] in the *International Journal of Game Theory*, which survived for several years.[53] [Roth: yeah] Later, there was a new managing editor who didn't want this anymore and it died down. It is a question whether it would have been good to have a journal maybe ten years earlier. I don't know. It worked out quite well as it was. And I'm not complaining about anything.

Chris Starmer Al, then Charlie.

Al Roth I'm glad that we didn't have the journal earlier, and what concerned me about that was the experience in game theory. I think that the *International Journal of Game Theory*[54] started too early, and in its early years was not a very good journal. Because *Econometrica* was publishing papers in game theory eagerly, and so good papers went to *Econometrica*. In some of the early issues of the *International Journal of Game Theory* you could see that they were really having trouble getting good papers.

And as game theory expanded so that there started to be lots of excellent papers that couldn't be published in the top economics journals, then *Games and Economic Behavior* formed, and I think it actually leap frogged over the *International Journal of Game Theory* because of the bad history that it was still suffering.

Reinhard Selten I don't think it was a bad history. It was just the personality of the editor. [one unidentified confirmatory sound] *Kalai* was a very good editor. [v. Winden: yeah] He approached people. He actively got people to [submit to the journal, if he heard interesting talks.]—I think it was just the personality of *Kalai*.

Jim Friedman And taste.

[53] There was a section on Games and Experiments directed by Selten from 1992 to 1995 then by Ron Harstad until 2000.

[54] The *International Journal of Game Theory* was founded by Oskar Morgenstern in 1971 (volume 1 appeared in 1971 but was actually completed in 1972) The first Managing Editor was Gerhard Schwodiauer who served until 1980 and was followed by *Anatol Rapoport* until 1984. The Editors were successively William Lucas, Cornell University volumes 9–12 (1980–1983); Reinhard Selten, University of Bielefeld, volumes 13–17 (1984–1988); Joachim Rosenmuller, IMW, Bielefeld, volumes 18–23 (1989–1994); Dov Samet, Tel Aviv University, volumes 24–29 (1995–2000); For details see: 2002. "Preface. International Journal of Game Theory, 1995-2001." *International Journal of Game Theory*, 31, 151–53.

Al Roth — *Kalai* deserves lots of credit, but in the early years of the *International Journal of Game Theory* there were papers with theorems that were incorrect. There were some real problems with some of the early issues.[55]

Reinhard Selten — There were some problems because the early editor, there was an editor there who was not very good. When I became the editor, it was for [a period of] five years, the quality increased a lot.

Al Roth — Absolutely. [group laughter] Absolutely.

Reinhard Selten — I must say I increased the quality, but just by the advice of the editors. I didn't take such an active role in doing things like Kalai. I was not an optimal editor, but nevertheless I worked on increasing the quality. But maybe more could have been done with this job.

Chris Starmer — Charlie. You wanted –

Charlie Plott — I guess, he is right [refers to Al Roth]. In these early days, I was not in favor of a specialized journal and argued against it for quite a long time. The problem was I felt that the major journals, the audience of economics, need to be exposed and might not be exposed if it was in its own little specialized journal. Whether it was a good decision or not I have no idea. But we see, in say, *Social Choice* where it split off and has a special journal. *Economic Design*, which is split off and has a special journal. I'm not so sure that either of those have been smart ideas from the point of view of trying to get those kinds of messages out to the general population.

Anyway, the story is we didn't start a specialized journal up until the '90s. And by that time, we had a lot of people. It was beginning to expand quite a bit. Plus, we ended up with two really first class editors who did a lot of work in getting that journal started.[56] How it would have turned out, whether it was a mistake or not, I don't know. But I know I was very skeptical of the benefits at the time.

---

[55] A precise reference has not been determined.

[56] The founding co-editors were Charlie Holt (1998-2004) and Arthur Schram (1998-2007). Subsequent co-editors were Tim Cason (2004-2009), Jordi Brandts (2007-2012), Jacob Goeree (2009-2014), David Cooper (2012-) and Charles Noussair (2014-).

## Lowering the Barrier to Entry: Editors and Handbooks

| | |
|---|---|
| Chris Starmer | Another feature on the landscape of the 1990s, it seems to me another interesting stage of development, perhaps, was the emergence of a series of books on experimental economics. *Davis* and Holt, Roth and Kagel, *John Hey*, and more.[57] I was wondering perhaps if either John or Al might tell us something about the development of your book. The idea for it. |
| John Kagel | The idea for the book I think came from the editor. Do you recall his name? [addresses Al Roth] No. The editor from *Princeton University Press* had been going – |
| Charlie Holt | *Jack Repcheck*, a great guy. |
| John Kagel | *Jack Repcheck*, yeah. He came to us and was pushing an idea of what about a handbook of experimental economics, something summarizing things.[XXXIX] Al and I were in Pittsburgh at the time, so he could get both of us at the same time. We decided that the context of the book would be to have chapters in which people would relate to a series of experiments. How one experiment led to an idea, there would be some conflict, and there would be a new experiment and how these things would be resolved. We wanted our authors to focus on series of experiments.<br>Then we went around individually to various authors and got surprisingly good responses from people like John Ledyard to give us a chapter on public goods, which was just perfect, and Charlie [Holt] on IO experiments, etc.[58] I guess it took us a couple of years before we accepted *Repcheck's*– |

---

[57] **Davis, Douglas D. and Charles A. Holt.** 1993. "Experimental Economics," Princeton University Press, 406–26, **Friedman, Daniel and Alessandra Cassar.** 2004. *Economics Lab: An Intensive Course in Experimental Economics*. London: Taylor & Francis Ltd, **Friedman, Daniel and Shyam Sunder.** 1994. *Experimental Methods: A Primer for Economists*. Cambridge [England]; New York: Cambridge University Press, **Hey, John Denis.** 1991. *Experiments in Economics*. Oxford, UK; Cambridge, USA: B. Blackwell, **Kagel, John H. and Alvin E. Roth.** 1995. *The Handbook of Experimental Economics*. Princeton, N.J.: Princeton University Press, **Roth, Alvin E.** ed. 1987c. *Laboratory Experimentation in Economics: Six Points of View*. Cambridge; New York: Cambridge University Press.

[58] The following eight chapters were contributed—1. Introduction to Experimental Economics by Alvin E. Roth; 2. Public Goods: A Survey of Experimental Research by John O. Ledyard; 3. Coordination Problems by Jack Ochs; 4. Bargaining Experiments by Alvin E. Roth; 5. Industrial Organization: A Survey of Laboratory Research by Charles A. Holt; 6. Experimental Asset Markets: A Survey by Shyam Sunder; 7. Auctions: A Survey of Experimental Research by John H. Kagel; 8. Individual Decision Making by Colin Camerer.

Al Roth — The funny thing is I don't remember it originated with *Repcheck* because I remember that we weren't sure at all what publisher we were going to go with. [Kagel: we were bargaining] And we interviewed a number of publishers. I thought that maybe *Repcheck* came into the picture after we already were underway, but I don't have a definite memory of that.

John Kagel — Now, my memory seems to be that *Repcheck* was promoting the idea, and then we did want to bargain with different publishers. Once we decided to put the enterprise together, we made him compete for it. And one of the key things that we bargained for were free copies [to be] sent to a great number of people. Princeton provided some 500 free copies to various people because we gave them a list of people we wanted them sent to, namely the editors of all the major journals to educate them, and they did that. And then we had some hassle from the series that put out the –

Al Roth — Elsevier.

John Kagel — Elsevier. Okay. And they wanted us not to use the title *The Handbook of Experimental Economics*. We said it wasn't copyrighted, so what is the problem? [some laughter]

Al Roth — Another thing we wanted from the publishers was that they should go quickly to a paperback version, because we wanted it to get into the hands of graduate students quickly. I remember that was a big motivation to lower the barrier to entry. [Hoffman: ehm] We had the feeling that there was starting to be enough experiments that you might be deterred from doing experiments because of the feeling that it would be hard to catch up with where you had to be in order to start, and we wanted to lower that barrier to entry.

John Kagel — Yeah. One of the bargaining points was on the price of the book, so we didn't want the Elsevier pricing policy, which was very steep, and I think we got a price of like $70.00 or something like that for the hard copy.

Al Roth — We wanted it to be in the hands of people, not just in libraries.

## Dissents on Method

John Kagel: Exactly. One of the issues that kept coming up was why not have a section on methodology per se? Our position was that methodology should be dealt with within the context in which it arose. It is hard to specify general principles. It is much easier to elucidate those in the context in which they arose, and so we asked all of the authors to focus on these methodological issues.

Chris Starmer: That seems quite an interesting thought because, even though economists had been doing experiments for some time there wasn't any established training for experimental economics. Did you have the impression that there was a shared sense of method from these different authors or competing ideas about what the methods should be?

Al Roth: We thought there were competing ideas that arose because of the context of particular experiments. Our feeling was that a good experiment is one that accounts for the plausible alternative hypotheses that you are dealing with, and that depends partly on what you are studying and even on who your audience is. We definitely didn't want to try to lay down the law and say these are the precepts for doing experiments.[59] We thought you could do anything you wanted if you reported it carefully and if it controlled for the relevant alternative hypotheses.

---

[59] The issue of precepts of economic experiments appeared earlier, most notably in Smith's seminal 1982 paper which built upon the few previous attempts "to articulate a "theory" of laboratory experiments in economics." **Smith, Vernon L.** 1982. "Microeconomic Systems as an Experimental Science." *The American Economic Review*, 72(5), 923–55. Smith defined five sufficient conditions for a microeconomic experiment—the precepts of nonsatiation, saliency, dominance, privacy, and parallelism. Other papers on this topics by experimentalists are: **Plott, Charles R.** 1979. "The Application of Laboratory Experimental Methods to Public Choice," C. S. Russell, *Collective Decision-Making: Applications from Public Choice Theory*. Baltimore, Md.: Johns Hopkins Press for Resources for the Future, 137–60, **Smith, Vernon L.** 1976. "Experimental Economics: Induced Value Theory." *The American Economic Review. Papers and Proceedings of the Eighty-eighth Annual Meeting of the American Economic Association*, 66(2), 274–79, **____.** 1982. "Microeconomic Systems as an Experimental Science." *The American Economic Review*, 72(5), 923–55, **Wilde, Louis L.** 1981. "On the Use of Laboratory Experiments in Economics," J. C. Pitt, *Philosophy in Economics*. Amsterdam: Reidel, 137–48.

For a philosophical discussion of precepts see: **Bardsley, Nick; Robin Cubitt; Graham Loomes; Peter Moffatt; Chris Starmer and Robert Sugden.** 2010. *Experimental Economics : Rethinking the Rules*. Princeton: Princeton University Press, **Guala, Francesco.** 2005. *The Methodology of Experimental Economics*. Cambridge; New York: Cambridge University Press.

| | |
|---|---|
| John Kagel | And the focus in terms of a series of experiments would be someone would do a particular experiment. They would have some alternative hypotheses in mind, and then it would occur to someone that maybe there was an alternative explanation or there was an element of the experimental design that did not permit certain sorts of behaviors and that this would lead to another experiment. And that is the context in which methodology should be discussed, because I think very much that this notion that a single experiment solves a problem is more or less of a fiction. Certainly, in my mind it is. And that raises this question of alternative explanations. |
| Chris Starmer | Did disagreements about methods surface at all through the meetings—German meetings, Tucson meetings, the Amsterdam meetings? |
| Reinhard Selten | There differences of opinions about methods were visible in the Amsterdam meetings. I remember this [pause] |
| John Kagel | I think it would more surface in the context of someone presenting a series of results or a result from an experiment, and then someone else saying I really don't think that is the explanation.[60] Okay. Here is my experiment. And now this, I believe, counters your explanation. Did it really do what it said it was doing and things like that? Then there would be controversy within that context rather than on general principles. |
| Charlie Plott | During this period you are talking about there is an amazing evolution that is taking place. Up until around 1985 or so, almost all experiments were done by hand. I even started a book, a rather large manuscript that I'd give to my students about here is a list of various and sundry instructions. Here is a list of various and sundry procedures that you would do if you were doing it by hand. But as we moved into the late 1980s, early 1990s, the technologies began to take over. So we see an evolution from experiments by hand into much more machine related things, and that once you start doing computerized experiments in one form or another, the methodologies begin to change rather dramatically, and still are changing as the technologies update. |

[60] **Editor**: Is there any particular *early* example that you recall? **John Kagel**: "One of the better early examples—although may be not early enough for you—would be an alternative explanation for bidding above the risk neutral Nash equilibrium first-price independent private value auctions—it is covered a bit in my 1995 chapter in the Handbook." **Roth, Alvin E.** 1995. "Bargaining Experiments," J. H. Kagel and A. E. Roth, *Handbook of Experimental Economics*. NJ: Princeton Univ. Press, For a historical treatment of this episode see **Svorenčík, Andrej.** 2015. "The Experimental Turn: A History of Experimental Economics," Ph.D. dissertation. University of Utrecht, pp. 193–231.

This discussion about methodology never really crystallized because it was always moving and always controversial. That is my take on it. We would always have this moving structure of methodology and how you do things because our equipment is changing all the time.

John Kagel — I have a little bit different take on the technology part. I remember doing first auction experiments by hand, and it was just very, very tedious.[61] And once we could figure out how to computerize it, it is a straightforward substitution of capital for labor, and in a lot of ways I thought it was.

Charlie Plott — I don't think so.

Chris Starmer — Charlie [Holt].

Charlie Holt — I just wanted to mention something about the book that I wrote with *Doug* [*Davis*]. Actually, it began as part of my work for the *Handbook* [*of Experimental Economics*]. The *Handbook* took a while to finish.[62] Even though this came out in '93, it was started after the draft of the *Handbook* chapter. I was talking to Kluwer and they had in mind like an $80.00 price and a low royalty rate. I called up *Jack Repcheck* at *Princeton* [*University Press*], and he flew down to Virginia the very next day with a contract. He was promising like a $35.00 price and double the royalty rate, and he was very eager. By the way, this is the editor who also coordinated the [volume on] winner's curse.[63] [Kagel: yeah] He was a key person in this.

But I thought writing a book for me was very helpful. It gave me a chance to read the things that Vernon and Charlie and Al and various people had done. And the great thing about a book though also, is that when you teach the class, you have something to rely on; makes it fun.

Chris Starmer — Frans.

---

[61] These pen and paper-based experiments were done around 1983 and eventually developed into two key papers on the winner's curse. **Kagel, John H.; Ronald M. Harstad and Dan Levin.** 1987. "Information Impact and Allocation Rules in Auctions with Affiliated Private Values: A Laboratory Study." *Econometrica*, 55(6), 1275–304, **Kagel, John H. and Dan Levin.** 1986. "The Winner's Curse and Public Information in Common Value Auctions." *The American Economic Review*, 76(5), 894–920.

[62] The *Handbook* was five years in the making from 1990 until 1995.

[63] **Kagel, John H. and Dan Levin.** 2002. *Common Value Auctions and the Winner's Curse*. Princeton, N.J.: Princeton University Press.

## What Constitutes an Observation? Part I

Frans v. Winden — I would like to come back to the methodological differences. I think they clearly surfaced, at least from my perspective. I was a newcomer and was observing what was happening and getting involved myself. I can remember at the Amsterdam meetings that there were some heated discussions, and they were related to three topics. First of all, what is an independent observation? What I remember is that people from the United States were more liberal in the sense that they applied parametric statistics whereas the Germans, especially of course Reinhard Selten and the group he influenced, were stricter on that. The problem is that if people are interacting, even if it is with random matching over rounds, they may affect each other. You should then take the group as a unit of observation. That was one of the issues that showed up and were every now and then, rather hotly discussed.

There were two other topics. The German perspective was more to use experiments not so much to show that standard theory seems okay or is doing a good job, but to go more into bounded rationality. It was a different belief behind the experiments that played a role in what people were doing and trying to show or hoped would come out.

Related to that was also that there was much more appreciation of doing experiments for exploration from the German perspective compared to the perspective of the people from the United States. The latter were more into hypothesis testing and taking established theories as a benchmark. I think these three things, as far as I can recollect, played a clear role in the discussions.

Chris Starmer — Okay. I had two hands from Vernon a moment ago.

Vernon Smith — Well, I just wanted to add to this conversation about publishers being very interested in publishing books and handbooks. Jim Friedman and I, Jim will remember, we proposed a collection of experimental papers in 1969. We sent it to *John Wiley* who had published both of the Siegel books. They said, "No, thank you. We didn't make any money on either of those."[64] [group laughter]

Jim Friedman — That is a sad memory.

---

[64] The publisher of the Siegel Fouraker volume *Bargaining and group decision making; experiments in bilateral monopoly* (1960) and the Fouraker Siegel volume *Bargaining Behavior* (1963) was McGraw-Hill.

Vernon Smith	I mentioned that to show you how much it changed from after '69 through the '70s and up into the '80s. A big change.

Chris Starmer	I have Betsy, Charlie, and then Al.

## Payment and Deception: Spoiling the Subject Pool or Spilling the Beans?

Betsy Hoffman	I wanted to add to the discussion on methodology. As experimental economics was separating itself from experimental psychology, one of the big methodological debates was [about] the importance of paying subjects, which psychologists still don't do. And the importance of not lying to subjects, which psychologists still do.

Experimental economists very early on coalesced around the ideas that it was very important to pay subjects, and that it was very important to be brutal, to be honest with subjects about what the experiment [is about; tell them] what exactly they were supposed to do in the experiment; and debriefing them later on; and that you would not be lying to subjects. I think that was a very important methodological discussion that experimental economists coalesced around early and really separated themselves from experimental psychologists.[65]

Chris Starmer	Reinhard, two hands.

Reinhard Selten	It is a strange thing since these principles do not go back to an economist but to a psychologist, namely *Sidney Siegel*.

Betsy Hoffman	That is true. You are correct.

Reinhard Selten	Siegel insisted on that, and that influenced experimental economists a lot more.

Betsy Hoffman	But not psychology.

Reinhard Selten	Yeah. But not psychology.

Chris Starmer	Two hands Vernon.

Vernon Smith	Most people, an awful lot of people were very much influenced by *Sid Siegel* and they had no idea they were.[XL] [Friedman: yeah] Now, *Sid* was very strong on paying the subjects, and you don't deceive them. Because when you do that, it just simply spoils the subject pool. That is what he worried about.

[65] **Editor**: Are you referring here to the 1970s? To what extent did these discussions appear in publications? **Betsy Hoffman**: "Yes, the 1970s. I don't think there was ever a publication, but it was discussed at meetings."

Chris Starmer — Okay. We have got a bit of a queue now, but starting with Charlie.

Charlie Holt — I just wanted to disagree a little bit with Frans about the American perspective on independent observations. If you look at Sidney Siegel's book on nonparametric statistics,[66] which was very influential for me, it takes all of the independence assumptions seriously. If you look at the chapter on methodology, which is at the end of *Davis* and Holt's—it is typical to put it [discussions of methodology and nonparametric statistics] at the end—that is basically a paraphrasing of what *Sidney Siegel* had talked about.[XLI]

On the other hand, I want to agree with Frans about the difference in the German perspective of let's not try to show the theory works, let's try to figure out where it doesn't work, what is going on and what is the behavioral factor. And I remember, when I worked with *Doug Davis*, coming from the theory side, I have always tried to set up an experiment where theory worked well. Then Doug had this intuitive idea. He would go change some parameter that didn't change the prediction but pulled the data away. And I asked, "Doug, why did you do that? Now, we have got to explain this." And he said, "Charlie, don't worry, it's just behavioral." Put that label on it, behavioral. Of course, we do a lot better than that now, and we try to actually explain the behavior.

Vernon Smith — Doug has got the right idea. Once you get it to work, you ask what makes it not work. [Holt: yeah] Otherwise, you are just spinning your wheels in the same rut. You are not going to learn much.

Al Roth — I wanted to pick up on Frans' point that different experimenters approached experiments with different points of view and different objectives. I think that continues to some extent. But I didn't see that as a national issue. Some time before the *Handbook*, I had a conference in 1985 in Pittsburgh that a lot of people came to. And it came out as a volume called *Laboratory Experimentation in Economics, Six Points Of View*.[67] The idea there was to try to elicit what people were trying to do with experiments. When we did edit the *Handbook*,

[66] It appeared in McGraw-Hill's series in psychology: **Siegel, Sidney.** 1956. *Nonparametric Statistics for the Behavioral Sciences*. New York: McGraw-Hill.

[67] The conference "Laboratory Experimentation in Economics" took place at the University of Pittsburgh on May 16–18, 1985. Participants included *Reinhard Selten*, Richard Thaler, Marc Knez and *Vernon Smith*, *John Kagel*, and *Charlie Plott*. **Roth, Alvin E.** ed. 1987c. *Laboratory Experimentation in Economics: Six Points of View*. Cambridge; New York: Cambridge University Press.

we talked a bit about how experiments were used from different points of view, we talked a little bit about speaking to theorists and searching for facts, and whispering in the ears of princes. Those are points of view from which you can approach the laboratory.[68]

Chris Starmer Jim.

Jim Friedman I have a question I would like to really place before those of you who know something about what psychologists actually do. It is the lying and the paying stuff. I think the paying is absolutely clear-cut because of obvious incentive issues. With respect to lying, it has always struck me that within the economic context we just don't have a reason to lie to the subjects. Generally, we want them to understand the circumstances they are in, and we want them to make what would appear to be intelligent decisions within that context. So we lay it out for them and let them loose.

Now, is it perhaps not the case with the psychologists that in some experiments, if they don't lie and they tell you the truth, they are spilling the beans and telling you how to behave. When in fact they want to elicit behavior that hasn't been poisoned. I don't know if that is right, and that is why I am putting this out.[XLII] You are an administrator, Betsy, and you have contact with lots of disciplines. I know Reinhard has always had psychology connections, and probably you have, too. [not clear whom Jim is addressing, likely the next speaker]

Al Roth I have just two quick comments about lying or deceiving subjects. First is just a funny anecdote, but the second is a contemporary issue that we are seeing in our lab. When I came to Harvard, I needed to set up a lab, and it was the first economist lab [there].[69] I had some meetings with research administrators and was explaining that it would be a no deception lab. The particular administrator was a psychologist, and she asked why. I talked about not poisoning the pool and a look of big relief came across her face. She said, "I knew that an economist couldn't have ethical objections to this method." [group laughter][XLIII]

---

[68] Roth classified experiments according to the kinds of dialogues they were a part of—as "Speaking to Theorists," "Searching for Facts," and "Whispering in the Ears of Princes"—for the first time at a symposium at the 1985 World Econometrics Congress. ____. 1987a. "Laboratory Experimentation in Economics," T. Bewley, *Advances in Economic Theory 1985 (Symposia of the 5th World Congress of the Econometric Society*. Cambridge University Press, 269–99. It was also preprinted in____. 1986. "Laboratory Experimentation in Economics." *Economics and Philosophy*, 2, 245–73.

[69] Roth moved to Harvard in 1998.

But in terms of the public good aspect, another non-deception issue has now started to surface. People have started to propose experiments where they say we are going to have people play a game, and then we are going to say something new is going to happen now. Other people are going to get to observe what you did in that game and perhaps penalize you if they wish. Something like performing judgment. That is not deception. They are not lying to them, but it has raised the question of if we started to do lots of experiments like that [Selten: Yes, this isn't deception], maybe that would also poison the pool because people would say when we come into the lab, we are going to do something, and then later people are going to look at it in ways that we weren't told about.

That is not a deception issue, but it is a public goods issue of keeping the people concentrated on the incentives we are trying to induce in the lab.

Jim Friedman — Well, is it necessary in psychology experiments to lie to subjects?

Reinard Selten — Na, ja.

Chris Starmer — I think that given the time we have left, I think the deception issue is an interesting one, but [Jim: Not for now] perhaps for dinner conversation. I'm conscious that I have had John [Kagel] on my list for some time. I don't know if you remember why you wanted to be on the list at this point.

John Kagel — No, I don't really.

John Ledyard — Just say something.

Chris Starmer — Don't feel forced. [laughter]

## What Constitutes an Observation? Part II

John Kagel — One of the things I did want to say, in response to what Frans was saying. I think there was a brief period of time where people would write a paper, and they would like grade theory. They gave it an A, B, C, or D. And I found those exercises extremely boring in the sense that if you gave it an A, why was it getting an A.[70] Was it the same mechanism that the theory had in the fundamental ideas underlying the theory, or was it perhaps something else. There was a very brief period there where you did have these papers that were just grading the

[70] Precise references for such papers have not been determined.

theory, but that passed and now it is much more that you have to take an extra step in your design to push the boundaries of the model in order to get the paper published in a good journal.

The other thing is this issue of what constitutes an observation, which I have called the German critique, is very much with us today. The disappointing thing to me is that, in some cases, it is doctrinal without just looking at what the residuals look like and what happens if you—there are ways of dealing with. For instance, take a group of people interacting in, say, a voting game.[XLIV] You have got three people who are voting. And you have got fifteen people in the room, so they are being reconstituted into different groups all the time. Now, is that a single observation? Well, you can test that. You can introduce statistics that will test that observation.[71]

Reinhard Selten What you test can only fail to refute the null hypothesis. You cannot really prove the null hypothesis that it doesn't matter.

Chris Starmer I think we are establishing that there is an enduring division here. [Kagel and Betsy laugh] I propose that we don't try to fight it out because—Charlie.

Charlie Plott I think that you are exactly right that this discussion is likely to degenerate because there is as many ideas about methodology here as there are people at this table. A lot of them are very strong. There is no resolution of these things. I disagree with the idea that one country has one view and one country has a different view. [Selten: No, no, that is what I would also disagree with] I see a whole ocean of different points of view, and I don't think that it has been resolved or it will be resolved.

Reinhard Selten I wouldn't say that it is a national question. [v. Winden: No[72]] Maybe it was in the beginning but over time the geographical concentration of the sides taken in the discussion has [become less and less pronounced.]

Charlie Plott Yeah.

Chris Starmer Vernon.

Vernon Smith Well, there are so many things you have to do to implement a test of theory, that are not part of the theory. There is just a huge gap there that people fill in in different ways, depending upon their experience. And I think it is very important though

[71] Kagel has published several voting experiments since 2005.

[72] **Editor**: What did you refer here to? **Frans van Winden**: "What I meant is that I thought/felt there was a geographical distinction at the time of the Amsterdam Workshop in Experimental Economics. Therefore, my "No" here. Today it is different."

for every experimentalist to challenge his own beliefs and his own experience and try to prove himself wrong. Because you know more about how you might be able to do that than anyone else.

## Signs of Success: And the Resources It Takes

Chris Starmer — We have got just under 10 minutes of this session left, and I have got one question that I'm keen to ask before we finish. So, if you don't mind, I will move to that. It comes with multiple parts. We have discussed at various points of today, reactions to experimental economics from the outside, conferences, and departments and reviewers of journals and so on. What I'm interested in is your view of how things have changed and where we have got to. And here really comes the question. Would you say that experimental economics has reached the point where it has fully established itself as part of the method proper of economics? And if so, what are the signals of it and what are the causes of it?

Al Roth — One of the signals that maybe it has established itself is we are starting to see some full-throated critiques. No one ever had to criticize experimental economics before it became an important part of economics. But now you can read articles saying that experimental economics is just as silly as other articles saying mathematical economics is.[73] [XLV] I figure that is a sign of success. [some laughter]

Betsy Hoffman — I think that is good. [Ledyard: He's right] That is very good. I'm not sure I can top that. I think one is that the number of people who—I mean it is now expected to be part of the tool kit of a well trained economist. Just as econometrics and theory are part of the tool kit of a well trained economist. [Addressing Ledyard] You don't think so?

John Ledyard — Go ahead. [Kagel laughs] I don't want to interrupt.[74]

Betsy Hoffman — I mean, maybe not as completely, but after a long hiatus when I was doing much too much administration, two years ago I started going to the *American Economic Association* meetings

---

[73] **Editor**: Which papers do you have in mind? **Al Roth:** "Here I have in mind for instance:" **Levitt, Steven D. and John A. List.** 2007. "What Do Laboratory Experiments Measuring Social Preferences Reveal About the Real World." *The Journal of Economic Perspectives*, 21(02), 153–74.

[74] Ledyard expands on this later on in this subsection.

again. And what I noted was that you can go to almost any session and somebody may throw in some experiments as part of the tool kit, as it were. To me that was an interesting experience. Also, seeing my own department, which has never established an experimental program. It had *Lisa Vesterlund* for a while. She has been the only other than me, but I don't really count because I was hired as an administrator. But two years ago, my department decided they had to establish one. That it was part of being a modern economics department to have an experimental group.

Those are the two or three data points that I have noticed that I think there is a big change from when I was a graduate student.

Chris Starmer John.

John Kagel I agree. I think there has been a big change. I don't know that it is something that every economist has to have in their tool kit. [Betsy: I agree] What is much more the case these days is that people will come in who are interested in a particular subject matter. There will be an experimenter there. They will have discussions over coffee and things and they will engage in a series of experiments. Then they will move off and continue to do what they were doing before in terms of that subject matter.

And it is not like they are being contaminated or that sort of thing.[75] I think that is the sign of acceptance. The other sign of acceptance is the number of people who primarily do experiments who are on the editorial boards of major journals. That can be counted.

Betsy Hoffman I was going to say that is maybe a better way of putting what I was trying to say that sort of every department expects to have somebody who knows about experiments and people go to those people and say would you help me. It has become more of a group—economics, in general, has become more of a group activity in part because of experiments.

Chris Starmer Charlie [Plott].

Charlie Plott I don't know how to answer the question how it is becoming accepted or not. But I know some of the tools and mechanisms by which it had become accepted. I think it also has to do with the reason you saw the big transformation in the late 1970s. It has to do with successful applications. Successful visible

[75] John Kagel refers here to Charlie Plott earlier comment about economists visiting Caltech for a semester or a year as Fairchild scholars and getting contaminated with the "experimental bug."

applications like the airport landing slots,[76] electricity, or more recently the FCC types of combinatorial auctions, trains in combinatorial auctions,[77] assignment problems. We have seen just time and time again starting in the late '70s that it isn't the science that is driving the popularity. It is the application that is derived from the science that becomes very visible.

And when somebody says what is it good for, you could report to—you just point to six or seven or eight different big results that have grown out of experimental economics, and that stops the conversation. It also markets it in the departments. If you look to see whether it is becoming accepted? Well, the more you see those types of big, visible things popping up, that is where the sale is coming. That is where the acceptance is coming, I think. And is it accepted? Personally, I think we have just scratched the surface of what is going to happen to this profession.

When you look at the level of experiments in chemistry or physics or other places, you are going to see that fifteen,

[76] In the wake of the 1978 *Airline Deregulation Act* Plott conducted the landing slot study jointly with David Grether and then graduate student Mark Isaac for the Civil Aeronautics Board. It concluded that the method of allocating slots at the four high-density airports by a slot committee process as inadequate and suggested using a sealed-bid one-price auction operating at regular, timely intervals with several other features such as computerized aftermarket with "block transaction". Their report was published as *Alternative Methods of Allocating Airport Slots: Performance and Evaluation.* Pasadena: Polinomics Research Laboratories, Inc., 1979. It got eventually published as **Grether, David M.; R. Mark Isaac and Charles R. Plott.** 1989 [1979]. *The Allocation of Scarce Resources : Experimental Economics and the Problem of Allocating Airport Slots.* Boulder: Westview Press.

Stephen Rassenti's thesis dealt with algorithms for combinatorial auctions allowing sales of blocks of slots: **Rassenti, Stephen J.; Vernon L. Smith and R. L. Bulfin.** 1982. "A Combinatorial Auction Mechanism for Airport Time Slot Allocation." *The Bell Journal of Economics*, 13 (2), 402–17, **Rassenti, Stephen J.; Vernon L. Smith and Bart J. Wilson.** 2002. "Using Experiments to Inform the Privatization/Deregulation Movement in Electricity." *CATO Journal*, 21, 515–44. **Charlie Holt**: "The only major FCC combinatorial auction that was actually implemented is Hierarchical Package Bidding for the 2008 Auction 72 in the C block, which was proposed and designed by experimental economists. **Goeree, Jacob K. and Charles A. Holt.** 2010. "Hierarchical Package Bidding: A Paper & Pencil Combinatorial Auction." *Games and Economic Behavior*, 70(1), 146–69. Experiments with Jacob Goree, John Ledyard, and then a graduate student Christopher Brunner that were conducted for the FCC were a key factor in their decision not to implement more complicated auctions. **Brunner, Christoph; Jacob K. Goeree; Charles A. Holt and John O. Ledyard.** 2010. "An Experimental Test of Flexible Combinatorial Spectrum Auction Formats." *American Economic Journal: Microeconomics*, 2(1), 39–57.

[77] **Brewer, Paul J. and Charles R. Plott.** 1996. "A Binary Conflict Ascending Price (Bicap) Mechanism for the Decentralized Allocation of the Right to Use Railroad Tracks." *International Journal of Industrial Organization*, 14(6), 857–86.

| | |
|---|---|
| | twenty, thirty years from now in economics. It hasn't even started in terms of the impact it is going to have. |
| Betsy Hoffman | [Ledyard inaudible discussion with Hoffman] It's right now. That is true. |
| Chris Starmer | John, did you want to add? |
| John Ledyard | I was going to just echo what Charlie is saying. My strong reaction to Betsy was that until we see every student in the first year graduate program having to run experiments, or even first year principles courses [taught with experiments] the way I did in physics and chemistry, [unfinished]. There is resistance [to that]. There are universities which still do not have experimentalists and have no plans to get experimentalists. |
| Al Roth | Most universities don't have experimentalists. |
| John Ledyard | Well, many. And the resistance comes from people who run regressions as much as it does from macro economists, policy types. It comes from a lot of different places, so I think things are a lot better than they were in 1970 or even in 1985, but there is a long way to go. [Hoffman: I would agree] We are not there yet. |
| Vernon Smith | Well, I think that is appropriate because it really takes a major commitment and deep development of human capital because you don't learn to run experiments by reading about experiments. [Ledyard: Yeah] Nobody does. That is true in physics, in chemistry, and all those fields. And that is a huge investment. You have got all this human capital, and it is operating knowledge and technique that have to be built up. I'm perfectly comfortable that it not be universal.[XLVI] |
| Betsy Hoffman | I'm not. [some laughter] |
| Vernon Smith | Well, you can have an acceptance of what comes out of experimental economics as a body of knowledge, and that is happening. But I would say that probably more universities ought to be doing it that aren't' now. But it takes a lot of resources. |
| Chris Starmer | Okay folks. Well, I think we have pretty much reached the end of the allotted time for this session. Thank you very much. We can break now for coffee, refreshments and preparation. |

# Chapter 4
# Funding

## Payment: Tightening Up the Structure of the Model

Chris Starmer: We will start the final session of the day on the homerun towards some drinks and dinner. One of our central topics for this session is related to funding, which we touched on in one or two ways. A characteristic feature of experimental economics is that experimentalists almost always pay task related incentives for participating in experiments. You have to raise some funds for those payments. And I want to explore that process a little bit.

But before getting into that, I would like to ask what seems to me to be a prior question, and that is where did the idea for having incentives tied to experiments come from to begin with in economics?

Reinhard Selten: It comes from *Sidney Siegel*.

Chris Starmer: Can you say more about that?

Reinhard Selten: Yes. *Sidney Siegel* had built a laboratory for psychology, learning experiments, and so on. He made this very strong point of that there should be incentives. [For instance,] in his dissertation[1] he got the result that behavior is more rational in a sense if you increase the incentives. It was later challenged because he didn't only change the size of the payoff, but also he introduced losses. So what he got might be an effect of loss aversion, but at least he interpreted it such a way [that

[1] **Siegel, Sidney.** 1953. "Certain Determinants and Correlates of Authoritarianism," *Dept. of Psychology*. Stanford University.



A. Svorenčík, H. Maas (eds.), *The Making of Experimental Economics*,
DOI 10.1007/978-3-319-20952-4_4

increased payoff size leads to more rational behavior]. He was very strongly attached to the idea that some monetary payments are absolutely necessary in order to get serious behavior of the subjects. And then he also [unfinished]

Chris Starmer Okay. We have got a long list, so let me come back on this. But I would like to bring some other people in as well. First, Vernon.

Vernon Smith I agree.[XLVII] But also *Ward Edwards* from the get-go was paying his subjects. What is interesting about Ward Edwards is that he trained a lot of these people that went into cognitive psychology, *Amos Tversky* was his student. So was, I can't think of his name[2] [pause]—and I find that pretty interesting that he trained them and yet they deviated from that tradition. *Ortmann* quotes me in here as I once asked *Amos Tversky* whatever happened to the tradition of *Sidney Siegel*, and he said, "You are it." That was not a compliment. [group laughter] That was a touché. [group laughter] He was putting me down. You have to understand the context, it was at Caltech.[XLVIII] And the psychologists were there, and we were having some problems with the methods they were using. And it was in that context that I said whatever happened to the tradition? And he says, "You are it. You continued a bad tradition." [Smith laughs]

Al Roth We have been talking about the positive impetus, but there was also a negative impetus that other economists, not experimenters, criticized experiments. And one of the ways how they sometimes criticized them was saying these are unmotivated choices. Andreas' in his paper mentioned the Friedman/Wallis critique[3] and things like that. But I think that was part of it. [It] is that one wanted to just make it clear that the subjects in the lab were making economic decisions. I think that was part of the pressure from the non-experimenters as well.[XLIX]

Jim Friedman Also the notion of paying people is in a way that is related to what they are supposed to be doing would seem self evidently correct to an economist's mind. You say you are going to be a business firm, and you are going to take home the profits you make or you are going to be in the decision making experiment, and you are going to get a money payoff that is strictly related in the right way to the pretend payoff. That would seem self-evidently correct to an economist mind.

---

[2] Other successful students of Ward Edwards were Paul Slovic and Sara Lichtenstein.

[3] **Wallis, W. Allen and Milton Friedman.** 1942. "The Empirical Derivation of Indifference Functions," O. Lange, F. McIntyre and T. Yntema, *Studies in Mathematical Economics and Econometrics in Memory of Henry Schultz.* Chicago: University of Chicago Press, 175–89.

| | |
|---|---|
| Chris Starmer | But some of the early experiments, for instance, I think some of the early market experiments reported in Vernon's '62 paper were done without incentives.[4] Can you speak of the time when you started to use incentives. It seemed that you thought you could do it without, but it was something that was brewing? |
| Vernon Smith | Well, one of the things I had to do when I developed the theory of induced valuation is explain why it is that sometimes things work when you don't pay.[5, L] In that [1962] paper there is one experiment where I paid the subjects. And the next generation of experiments I did, I was paying them.[6] In fact, I made some comparisons showing that it didn't work. There are some pathological cases where if you don't pay, it doesn't work. I didn't have funding yet, but it didn't cost much, so I paid for it myself. |
| Chris Starmer | Was your move in that direction driven by what you were observing in the experiments? |
| Vernon Smith | Oh yeah. And my first funding was in '62 from NSF. And after that—see that was the year that first paper was published. After that, the subjects were always paid. |
| Chris Starmer | There is a couple more people waiting to get on this. John [Kagel] and then Charlie. |
| John Kagel | The Wallace/Friedman critique that Al mentioned and that Andreas covered has been very much with us. It is an empirical question of what happens if you don't pay subjects and how much do you have to pay them and things like that. But it was the impetus of the Wallace/Friedman critique, at least for me. |
| Charlie Plott | Well, I don't know about the rest of these guys, but I never believe any of these things any one of these guys tells me about data or experiments. I have to do it myself. Vernon was telling me [about] paying his subjects, but of course I didn't really believe that it was necessary. I had run a whole series of experiments [including] some with Vernon.[LI] In markets, if |

[4] Smith's 1962 paper states in its first footnote: "The experiments on which this report is based have been performed over a six-year period beginning in 1955. They are part of a continuing study, in which the next phase is to include experimentation with monetary payoffs and more complicated experimental designs to which passing references are made here and there in the present report." The issue of payoffs is repeatedly raised in **Smith, Vernon L.** 1962c. "An Experimental Study of Competitive Market Behavior." *The Journal of Political Economy*, 70(2), 111–37.

[5] ____. 1976. "Experimental Economics: Induced Value Theory." *The American Economic Review. Papers and Proceedings of the Eighty-eighth Annual Meeting of the American Economic Association*, 66(2), 274–79, ____. 1973. "Notes on Some Literature in Experimental Economics." *Social Science Working Paper, California Institute of Technology*, 21, 1–27.

[6] That experiment was conducted after the manuscript of the 1962 was written and is reported in the paper's footnote 9.

you don't pay the subjects a commission, that last unit is not traded. It is easy to see it. If you don't pay them anything at all, you discover that sometimes it will converge, sometimes it won't.[7] [Experiments in which subjects are not operating with controlled incentives] are not reliable.

In committee experiments, if you don't pay them, it turns out fairness just takes over like crazy. You will see the system go from a core/equilibrium to something that is quite different and sometimes quite arbitrary. So, the thing we discovered was that if you don't pay them, all of a sudden you start getting a wide variety of outcomes that we thought are inexplicable. When you do pay them, all of a sudden the structure of the models tightens up, and the models work. So, in fact, we even pushed payments up to $500.00 or $1,000.00 to see if the levels of the payment made a difference, and it turns out no.[8] Getting it really high doesn't make much difference.

This is again in the very early '70s when I adopted this [methodology of paying subjects]. It wasn't because somebody told me to do it. It was because the data came out that way.

Chris Starmer John [Kagel].

John Kagel Levels of payment will occasionally have an impact. I liken it to a situation where you are teaching a course and you say this is going to be on the test. Then the students really listen up. I think it has to do with where they have to think further ahead. We had one experiment in one of our signaling games where we increased [the payoffs significantly.] We were able to do this in China, at least back then.[9] We normalized things compared to the cost of living or purchasing power in the U.S. and we increased it by a factor of five.

[7] **Miller, Ross M.; Charles R. Plott and Vernon L. Smith.** 1977. "Intertemporal Competitive Equilibrium: An Empirical Study of Speculation." *The Quarterly Journal of Economics*, 91(4), 599–624.

[8] The issue of the size of payments has received considerable attention, especially in the context of the ultimatum game and other bargaining situation in low income countries see for instance: **Roth, Alvin E.; Vesna Prasnikar; Masahiro Okuno-Fujiwara and Shmuel Zamir.** 1991. "Bargaining and Market Behavior in Jerusalem, Ljubljana, Pittsburgh, and Tokyo: An Experimental Study." *The American Economic Review*, 81(5), 1068–95, **Slonim, Robert and Alvin E. Roth.** 1998. "Learning in High Stakes Ultimatum Games: An Experiment in the Slovak Republic." *Econometrica*, 66(3), 569–96. **Hoffman, Elizabeth; Kevin A. McCabe and Vernon L. Smith.** 1996. "On Expectations and the Monetary Stakes in Ultimatum Games." *International Journal of Game Theory*, 25(3), 289–301. For a definition of the ultimatum game, see Chapter 5, Footnote 2.

[9] **Cooper, David J.; John H. Kagel; Wei Lo and Qing Liang Gu.** 1999. "Gaming against Managers in Incentive Systems: Experimental Results with Chinese Students and Chinese Managers." *American Economic Review*, 89(4), 781–804.

And what we did find was that the initial behavior was much closer to equilibrium. They were thinking much further ahead. But the entire path was essentially the same. But that is very unusual. That was an increase of a factor of five, which was quite large.

## Paying the Subjects: Tax Money Spent on Frivolous Things

Chris Starmer: Can we move towards the question of, if we are using incentives, where was the funding coming from? Where did people start getting funding for paying participants? Reinhard.

Reinhard Selten: In the beginning, when I still worked in Frankfurt, *Sauermann* had this special money and we could fund payments from that. But then when I was for three years in Berlin I could not get any funding. I did some experiments, paid by my own private money.[10] And then when I took the position in Bielefeld, verbally they told me I could use 10,000.00 DM a year from the general budget. But when I tried to do this later, they told me it is not legally possible. It is not legally possible because you cannot pay people for playing games. If they want to play games, they have to bring their own money and so on. [some laughter]

John Ledyard: It is okay to run a gambling hall but not to run experiments.

Reinhard Selten: I didn't do anything with this, because I am not a very active person. But then I wrote them a letter and said, "I'm known as an experimenter and I have to explain to the public and scientific journals why I can't pay money to the subjects according to their success." And I wanted to warn them that I am now going to do this.

Chris Starmer: Do you remember when you did this?

Reinhard Selten: No. Then they asked me to come, and then they said, "Well, we will give you 3,000.00 [DM]." The reason is that all these officials, at least in Germany but maybe also everywhere, the worst that can happen to them is that they violate the law. [Ledyard: oh yeah] That is very bad for them. They want to avoid risk. Now, you have to put them before another risk.

[10] These experiments remained unpublished. Selten was a full Professor at the Department of Economics of the Free University of Berlin, Germany, in 1969–72. In 1972 he moved to University of Bielefeld.

| | |
|---|---|
| | [some laughter] The other risk would be that they would be getting bad press. [some laughter] Therefore, I did this, and then I got 3,000.00 DM. In the worst case, I would cover it [subject payments] from my own money. This university administration really knew that it would get bad press. And then later, we even used more from the general budget because nobody objected. |
| | We just used more in Bielefeld. Later when I came to Bonn, I got it in writing that I can use money in this way.[11] Otherwise we also couldn't use the money from the research grant in this way [for subject payments] as long as they took the point of view that it is illegal to pay subjects according to their success in the experiment. |
| Chris Starmer | That is a serious constraint for it. |
| Reinhard Selten | This is a very serious constraint, but it can be overcome. But it takes some effort. |
| Chris Starmer | John. |
| John Kagel | I think within the United States, *Dan Newlon* at *National Science Foundation* is critical. I think Charlie [Plott] had a huge influence on *Dan* as well. And – |
| Chris Starmer | Can you tell us a bit more about it? |
| John Kagel | The *National Science Foundation* wound up supporting a number of laboratories.[LII] When we first set up an animal lab, we got our money from the NSF. And for most of the people at least from the States around this room, we are getting money from the *National Science Foundation* effective when Dan became the head of the *Economics Division* of NSF. And he was very supportive.[LIII] |
| Chris Starmer | Do you know when was that? |
| Betsy Hoffman | It was about 1975.[12] |
| John Ledyard | I was on the panel[13] in '78/'79. *Jim Blackman* was still a very important figure, and *Dan* was his assistant essentially at that point. I don't think he was quite yet the head of the econ program. But he was very influential, and *Dan* swam upstream against a number of trends in economics at that point. He supported public choice, he supported theory, he supported experiments and was very helpful.[LIV] |
| Chris Starmer | Charlie. |

[11] Selten moved to Bonn in 1984.

[12] Newlon came to NSF in 1974 as Associate Director for the Economics Program. He became Director of the Economics Program in 1980.

[13] Ledyard served at the Advisory Panel for the Economics Program during 1978–1980.

Charlie Plott — I can give you an interesting event dealing with this. When I first started, I think I had my NSF grant in maybe 1970.[14] And it was for doing social choice theory types of things. I became interested in the committees and started using some of that money for social choice for committee experiments, and in particular, voting agenda experiments. I was doing an experiment at UCLA. I just finished the experiment, and I was paying the people. And in the door walked *Jim Blackman*, the NSF officer who happened to be at UCLA talking to somebody else. He decided he would come over to Caltech. He asked my secretary where I was. He said he's on UCLA and told him.

There is Jim, I'm passing out his money [Smith referring to Charlie: he's spending his money] for something that he hadn't [agreed to] paid for. We sat down, and I explained to him in detail what I thought the advantages of laboratory experiments were, why this experiment bore on that. In fact, we spent much of the afternoon talking about what that meant and how you carried on the conversation with the other sciences if you were—apparently, *Blackman* was quite supportive after that. And that got renewed. Then, of course, following that *Dan Newlon* was supportive of everyone. He is also a University of Virginia graduate, so we knew him quite well.[15]

But following that though, in terms of funding, my own funding came from a whole series of applied projects. We did the first applied project with the barge problem that was funded by the *Department of Transportation*.[16]

Chris Starmer — Could I put the applied on hold for a minute. We are going to come back to that, so there will definitely be an opportunity.

Charlie Plott — Okay, that is fine. In terms of funding, there is a whole series of funding that comes from those sources.

Chris Starmer — Yeah. I want to talk to that, but just putting them on hold for a moment. Betsy.

Betsy Hoffman — I just wanted to add a little bit on *Dan Newlon*. What I didn't say before is that I was a subject in the Hong and Plott

[14] Plott's first NSF grant, GS-36214, titled Political Economic Decision Processes with $59,800 awarded covered the period 11/72–11/73.

[15] Plott graduated from Virginia in 1965, Newlon in 1970. University of Virginia was a center of public choice research with James Buchanan and Gordon Tullock on faculty.

[16] **Hong, James T. and Charles R. Plott.** 1982. "Rate Filing Policies for Inland Water Transportation: An Experimental Approach." *The Bell Journal of Economics*, 13(1), 1–19.

| | |
|---|---|
| | experiment.[17] [Plott: oh, that's true] I had the experience of actually being a subject and then later took his class. And even in the class, he insisted that we pay subjects for the initial experiments that we did. When *Matt Spitzer* and I put in our first grant proposal in 1980, we got the usual really bad referee reports. And *Dan* just – |
| Chris Starmer | Can I just ask were subject payments coming through as part of what people were complaining about? |
| Betsy Hoffman | I think they were complaining about the whole research agenda. This was the agenda to study the Coase theorem. The whole idea that you could actually test the Coase theorem was anathema to the reviewers. But *Dan*, I remember, gave us $25,000.00. We took no summer support ourselves. We took no payment. We hired no graduate students. We did all the work ourselves. And out of $25,000.00, we wrote six papers and won an international prize for the research.[18] But it was *Dan Newlon* just funneling a small amount of money, just enough money our way to pay our subjects. |
| Chris Starmer | Vernon then John. |
| Vernon Smith | I just wanted to say Charlie [Plott] had mentioned earlier that the NSF featured some experimental work in order to build up constituency within the NSF. And I remember one year, our grant was featured. I think, Charlie [Plott], you had that happened once. He rotated it and he used that to basically help to build support within NSF, not only just for experimental economics, but for the Economics Division. That they are doing something useful.[19] |
| John Ledyard | I was going to say two quick points. Again, as a result of my experience on the NSF panel. There was sizeable resistance to experimental payments among most of the people I served with on the NSF panel, many of whose names have gone on to become very famous.[20] This was budget money being spent on frivolous things as opposed to serious data gathering like panel data. It looked like it was going to be costly and it was |

[17] See footnote 185 for reference.

[18] Ronald H. Coase Prize for excellence in the study of law and economics (jointly with Matthew Spitzer), given by the University of Chicago Law School, University of Chicago Law School, 1986.

[19] **Blackmann, J.** 1978. "Experimental Studies of Choice in Economics." *NSF Program Report*, 1 (3 (June)), pp. 7–14.

[20] **Editor**: Who were those panel members? **Ledyard**: They included Michael Spence, James Heckman, Robert Barro (a macroeconomist), Arthur Goldberger (an econometrician)." The first two went on to become Nobel Laureates.

going to take away from other things. There was a lot of resistance and it was actually one of the things that made it harder to get them to approve experimental work. And as you say, *Dan* swam against that and ignored it. But it was intellectually a difficult fight.

Second, it should be pointed out that a lot of people in the '70s and '80s who started getting into experimental work, which wasn't very widely done and [wasn't] very hard, in many cases when applying for grants, it was important that you had some pilot experiments done ahead of time [of the application]. So either you got it out of your own pocket or it turned out that both Vernon and Charlie were very good about passing out small amounts of money from their grants to help people run those early experiments that then led to [experimental research and getting grants]. They broke the waves in very important ways.

Betsy Hoffman I actually got a small grant from Northwestern [University]. Their vice president for research had a small faculty start up grant. *Matthew* [*Spitzer*] and I were both at Northwestern, and we wrote a proposal. And so that is how we did our –

Chris Starmer When was that, Betsy?

Betsy Hoffman That would have been about '79 probably. It is listed on my vita. It can probably tell you exactly what date it is.[21] Yes, 1979.

Chris Starmer Jim.

Jim Friedman When I was starting my dissertation in 1961/1962, I wanted to run an experiment and there was the issue of payments for subjects. I didn't need to hire anybody. I just needed to pay subjects. And so I went to *Tjalling Koopmans* who was the director of Cowles Foundation at Yale at the time. I said, "I wonder if I could get some money, and this is what I want to do." He said, "How much?" And the number of like $500.00 or $800.00 sticks in my mind, but I don't swear to it. Whatever it was I said, he said, "Okay." Then I did my experiments, I finished my dissertation, and got a degree. Then afterward, I was staying at Yale as an assistant professor and I was part of a Cowles staff as well. The next thing I wanted to do in research was I wanted to run an experiment.

*Tjalling* [*Koopmans*] was probably still director, but I don't remember for sure. I had the same conversation over

[21] Northwestern University Law School Grant awarded to Matthew Spitzer for joint research, 1980–1981. Northwestern University Grants Committee Grant, 1979–1980.

again, but I think it was a slightly higher price, and then did that. And then the next thing [the experiment] I did. At that point, I thought that after doing that experiment that I was going to turn my total attention to doing theory. But *Austin Hoggatt* a couple of years later convinced me that he and I should do the work that you are very familiar with.[22] That he and I should do something together. For that we went to the *National Science Foundation.* We put in a pair of coordinated grants from the two different institutions. I was at that time in Rochester.

Both grants specified the experimental plans that we had and my application also talked about some totally independent theoretical work that I wanted to do. We got the research grant and it was renewed at a later point for another several years. I don't know who was running the shop at NSF at that point in 1969 when the first grant was awarded. It was absolutely clear from the paperwork that came back that the theoretical work that was part of that grant, which turned out to be the absolutely most prominent thing that I did in my life professionally,[23] that that was of no account and the experimental work [some laughter]

Vernon Smith It was all that mattered.

Jim Friedman Yes. And the experimental work is what got the grant. [group laughter]

John Ledyard Wow, that is interesting.

Frans v. Winden That is amazing.

Jim Friedman There is a lot of noise in life.

Betsy Hoffman Oh yes.

Chris Starmer Okay. There is a little bit of a queue starting with Al, then John and then Charlie.

[22] Publications and grant details mentioned in Chapter 3, Footnote 5. 1969–71 NSF grant: "Theoretical Research and Collaborative Experimental Research on Micro-economic Games" 1972–74 NSF grant: "Theoretical Research in Game Theory, Oligopoly, and Individual Behavior"—joint but separate with Hoggatt.

[23] **Friedman, James W.** 1971. "A Non-Cooperative Equilibrium for Supergames." *The Review of Economic Studies*, 38(1), 1–12, ____. 1973. "A Non-Cooperative Equilibrium for Supergames: A Correction." *The Review of Economic Studies*, 40(3), 435. Friedman in this paper provided perhaps the first folk theorem type result.

## How to Fund an Economics Lab?

**Al Roth** There are two streams of finance we have to worry about and one is experimental subject payments, but the other is running the labs. [Hoffman: Yes] Mostly, in my experience, NSF grants have not been sufficient to run the labs, not even close. NSF doesn't have that kind of money. While I have gotten subject payments from NSF, I have gotten institutional support for the labs. At Illinois, we used existing facilities that we could borrow. At Pittsburgh, *Kevin Sontheimer* went and got a grant from the *Scaife Foundation* to buy an initial set of machines. At some point, there was some NSF facilities money as well, but the lab at Harvard [unfinished]

My first offer from Harvard was just from the economics department and FAS.[24] We didn't go on that one, and the FAS Dean was concerned about the cost and didn't see it right away, and so we stayed in Pittsburgh.[25] The subsequent offer was joint with economics [department] and HBS.[26] That is the one that I finally took. And when I talked to the HBS Dean, I said there is a lab involved, and when I previously had this discussion about the lab, that was an issue. And he said to me, "You know anything we decide we want to do, we can afford." [some laughter]

**John Ledyard** That is HBS and it's true.

**Al Roth** Right. [some laughter] Again, we get institutional support for the lab, for staff, for things like that. And that is different than say in chemistry.

The reason Deans have trouble with this sometimes and maybe this is an issue in the penetration of experimental economics into departments is if someone says to you I'm going to hire a chemist, and I want you, the Dean, to support his lab forever. Then you say what is wrong with that? He must be a bad chemist because good chemists can get the large grants they need, and that is just not true for economists. [supportive sounds by Ledyard and van Winden]

---

[24] FAS stands for Harvard's Faculty of Arts and Sciences. Roth declined this offer in 1996.

[25] **Editor**: What exactly did the dean not understand? **Al Roth**: "The FAS Dean told me there wasn't room in Littauer at that time, but that if I came to Harvard I could eventually have a lab once the government department was relocated to a new building, freeing up space in Littauer." Littauer building houses the Harvard Economics Department.

[26] HBS stands for the Harvard Business School. Roth accepted the joint offer in 1998.

John Kagel: Well, in terms of what happened at Pittsburgh when I was hired there.[27] Part of my going there was they had to provide a lab. I mean, that was the motivation for *Kevin* [*Sontheimer*] getting this money together. He also provided funds for a lab administrator. [Roth: yeah] At the University of Houston,[28] when I first started doing experiments with people, we used internal funds. Here was another case where the physical scientists—Houston is the oil capital of the U.S.—had an institute for oil research.[29] Doing mineral rights common value auctions was just right up their alley [Smith: ehm], and they were so delighted to be able to support something in social sciences and gave us money right away.

Chris Starmer: Charlie [Holt].

Charlie Holt: I remember my first NSF grant, which was related to auction theory. The panel came back and said, "By the way, why don't you do some experiments." [some laughter] I never thought of that.[30] In my second grant proposal I included some experiments, and it was funded.[31] My third [submitted grant] had no experiments, it was not funded, and I started to get the message. The other thing about funding, I was talking to Stephen Rassenti this morning, and he was pointing out one of the problems, and Vernon found this out. When you have funding for a lab and administrators want to go after it because it is very tempting. If it is money for subjects and stuff like that, they can grab it. If it is built into somebody's salary, they can't really take it away. [Roth: yeah] That soft support is really valuable and also vulnerable. [Smith: ehm]

Charlie Plott: Well, in terms of funding. I will just outline some of mine because I have never had university support ever, except that they gave me my space. Everything that I have had came through either NSF, and I have had NSF grants I guess starting in 1972, every year up until very recently, then I haven't applied.[32] The other parts of it came from the *Department of Transportation*—these are these applied

[27] Kagel moved from Houston to Pittsburg in 1988.

[28] Kagel spent the years 1982–1988 in Houston.

[29] Energy Laboratory, University of Houston.

[30] That was Holt's NSF grant titled *Signaling Auction Markets* for the period 1980–82.

[31] That was Holt's NSF grant titled *Experimental Studies of Industrial Organization Theories* for the period 1983–85.

[32] Plott's last NSF grant ended in 2008. It was a joint grant with William Zame titled *Collaborative Research: The Evolution of Prices and Allocations in Markets: Theory and Experiment* 2003–2008.

project—*Federal Trade Commission*, *Civil Aeronautics Board*, the FTC again, a big grant with NASA, which basically created some of the infrastructure, [Ledyard: gave you your [lab] room] and that really got us started as a lab. That came through the space station applied project. That was the first time we had any money at all for an administrator or technicians.

We never had that ever prior to that. And then that was followed by large grants from *General Motors Corporation*, *Bradley Foundation*, and an NSF instrumentation grant.[33] Starting around 1990, I started seeing money big enough to actually pay people aside from myself and just doing the grants. Up until around 1990 or 1985, this was really—mine was all a hand to mouth, only enough money to pay subjects. Not even enough money to pay graduate students because NSF didn't have it.

Chris Starmer: Reinhard.

Reinhard Selten: When I was at the University of Bielefeld, I had no lab. I could occasionally do experiments. I could do experiments without any lab. But I didn't have a lab. When I got an offer from Bonn, then I thought what should I ask for and I asked for a lab. In Bielefeld they were also bargaining and in Bonn they gave me enough money, and I also wanted to have three years to use the money. They gave me 120,000.00 DM for three years. Personnel was not a problem because my predecessor, Wilhelm Krelle, had four assistants and I could get them.[LV] I had enough people for the lab, but it would have to be built up. Computers at that time were still quite expensive. In the first year they were much more expensive than in the second and the third. It went down quickly and it was very good that we had this for three years.

But then there were still administrative difficulties because at that time there was at the university a data commission that had to approve all use of computers. The data commission didn't want us to build up a laboratory, which was not used for students, but exclusively used for experiments. I had some trouble for some time. But then I knew somebody who had influence on this, and he picked up the telephone and afterwards everything was done. Sometimes administrative difficulties are very serious.

Chris Starmer: Frans. You have been waiting sometime in the queue. I hope you remember that.

[33] See Chapter 6, Footnote 12 for details.

Frans v. Winden For setting up the lab and for our first experiments for a period of five to six years, we had this pioneer grant from the *Dutch Science Foundation*, which was extremely helpful to set up the research group. But then after the expiration, it got very difficult because the faculty was not willing to pay for subjects. To get it from NWO was also problematic because if you wrote a proposal, you had to submit it to a committee that was composed of economists. As economists were not into experimental economics, they were basically not very eager to pay for that. We couldn't rely on that and what it boiled down to was that we had to find contract research money in order to continue the existence of the research group and the lab.

Later it turned out that the faculty was willing to match the renovation costs of the lab, but that was the only thing. We had to rely on contract research money. Fortunately, we could rely on contract research that we liked, that was interesting to us from a fundamental research point of view. We did research, for example, related to the flower auction, railroads, and spectrum auctions.[34] These were all topics that we were interested in. Later on, also the European Union came in, and we could rely on some money from there. And gradually also the *Dutch Science Foundation* accepted more proposals where you asked for subject payment money. But it was difficult.

## Contract Money: The Best Kind of Money?

Chris Starmer Okay. Can we turn to the theme of applied and contract research because for a number of people here that has been an important dimension of the work. Steve and Charlie [Plott]

[34] Other topics included coordination in payment systems, committee decision making, and tax reform. 1995-97: research project for the Flower Auction Aalsmeer titled "The role of information in the flower auction: an experimental study." 1996: research project for the Radiocommunication Agency of the Dutch Ministry for Transportation and Waterworks) titled "The auctioning of ether frequencies by the government." 1998-99: research project for the Dutch Ministry of Social Affairs and Employment titled "An experimental study of the Van Elswijk Plan," that examined an alternative tax system for financing unemployment benefits. See Footnote 237 for additional context. 2003-04: Dutch Central Bank: research project "Impact of rotation on committee decision making." Published as: R. Bosman, P. Maier, V. Sadiraj, and F. van Winden, "Let me vote! An experimental study of vote rotation in committees, Journal of Economic Behavior & Organization, Vol. 96, pp. 32–47. 2007-08: research project for the Dutch Central Bank titled "An experimental study of high-valued payment systems."

have just been raising this. Perhaps you too, John, others maybe, too. Could people say a bit about how they got into that line of work? What drew them into it? Charlie.

Charlie Plott: My first contract was with the *Department of Transportation* for the barge project. And it was because I knew the chief counsel for the *Department of Transportation*.[35] He was interested in an ongoing controversy.[36] After that, we got funding from the *Federal Trade Commission* which was interested in a project dealing with information advertising in the professions, and they came to me and told me about the problem and so I applied.[37] Also, I got funding from the *Federal Trade Commission* as an expert witness in a federal case called the Ethyl case. And so that generated a whole series of research funding.[38]

That was followed by the CAB [*the Civil Aeronautics Board*] who funded our research to actually look over changing the airport access landing thing.[39] It has just been a whole series of small contracts. How do you get into those? I have no idea how to do it. I did not chase the contracts. Usually, in these cases, it has been some economist who is high up, has a problem, and he thinks experiments might be useful [to solve them]. The contracts always come from them. They have always been correct, by the way. They have always been really fascinating path breaking type of projects. But they have also been absolutely crucial in keeping my lab running because without them, I couldn't have done it.

Chris Starmer: Steve, you were nodding there.

Stephen Rassenti: Yeah. I think it is the best kind of money because usually the people that come to you have something in mind, and they are willing and strongly supportive of the work you are going to do for them. Sometimes you go out looking for this kind of money. I remember Vernon and I one time went to visit

---

[35] John Snow, the later Secretary of Treasury in the years 2003–6. Like Charlie Plott John Snow received his Ph.D. in economics from the University of Virginia in 1965.

[36] This is the Hong and Plott (1982) paper mentioned earlier several times.

[37] **Plott, Charles R.** 1981. "Theories of Industrial Organization as Explanations of Experimental Market Behavior," S. C. Salop, *Strategy, Predation, and Antitrust Analysis*. Federal Trade Commission. **Plott, Charles R. and L. L. Wilde.** 1982. "Professional Diagnosis Vs. Self-Diagnosis: An Experimental Examination of Some Special Features of Markets with Uncertainty," V. L. Smith, *Research in Experimental Economics*. Greenwich, Conn.: JAI Press.

[38] **Grether, David M. and Charles R. Plott.** 1984. "The Effects of Market Practices in Oligopolistic Markets: An Experimental Examination of the Ethyl Case." *Economic Inquiry*, 22(4), 479–507.

[39] See Chap. 3, Footnote 76 for references and further context.

| | |
|---|---|
| | *Southern California Edison*[40] concerning some of the work we were doing in electric power. At a certain point, we were sitting with the executive of the company and their legal staff. And one of the lawyers stood up. It was at a meeting like this [witness seminar], and he put his hands on the table. He said, "You realize these guys are talking about doing science not consulting." They asked Vernon and me to leave the room, and we never got a cent from *Southern California Edison*. [group laughter] |
| Vernon Smith | The other thing the lawyer added is we can't control the outcome. |
| Stephen Rassenti | Yeah. And we always insisted on [having the right to publish anything we discovered in an academic journal.] |
| Vernon Smith | And I remember, they all looked at me. I said, Yeah, that is why you want to do it, and you will learn something." [Rassenti: yeah] Interestingly, the management was on board, [Rassenti: yes] but they were turned around by the legal.[41] |
| Stephen Rassenti | But in other cases, the legal departments don't get in the way. We would have this kind of work in this industry from other sources over the last ten years. And it is usually somebody, as Charlie mentioned, who is eager. Usually an economist who works for a firm like this who gives you a call and invites you to come and take a look at a problem that they are having and asks you whether you can do anything in this area. Now, we have also had interest from government people, too. For example, we are working for the government of Singapore right now, but the same thing happens.[42] It is usually someone in the organization who has been touched by experimental economics somewhere along the line who will give you a call. |
| Chris Starmer | And of those cold calls that you get, what proportion of those turn into contracts? |
| Stephen Rassenti | They almost always do. That has been my experience. If they are eager enough to call you, you can usually work out some [arrangement]—because for most of these firms or governmental organizations, the budget required to support one of these investigations is much less than they spend on |

[40] Southern California Edison was the primary electricity supply company for much of Southern California.

[41] This example is reported in more accurate detail in **Smith, Vernon L.** 2008a. *Discovery—a Memoir*. Bloomington, IN: AuthorHouse. pp. 302–6.

[42] This project concerned wholesale market price manipulation in the Singapore electricity grid, and the value of vesting contracts as a tool to control price volatility.

| | |
|---|---|
| | many other cute types of endeavors that they undertake. For example, I remember Vernon and I, we worked for the—what was the name of the agency for the Australian government[43] that was responsible for putting the new wholesale market plan into place? |
| Vernon Smith | The *National Grid Management Council.* |
| Stephen Rassenti | When Australia was thinking of deregulating the power industry and going to a new regime, they actually did some of their own paper experiments at a cost of several million dollars |
| Vernon Smith | Two million. |
| Stephen Rassenti | and got very unreliable results.[44] |
| Vernon Smith | They were useless. |
| Stephen Rassenti | Useless results. Basically people were conferring with one another. They were not paid. It cost a lot to execute them because they hired people to execute these experiments and to analyze the data, but they had a bunch of people basically in various parts of the industry and asked them if this were the situation, how much would you offer to sell your power for? They went to—what is the name of the firm in Sydney that was controlling all of the New South Wales' power? |
| Vernon Smith | *Prospect Electricity.* |
| Stephen Rassenti | Yeah. And they said, "Well, we will pretend to divide this firm up into three pieces," and the guys are sitting in the office next to each other. They handed them all pieces of paper and asked them what would you do if you were doing this here. So a couple of million dollars later, they decided they might go a different route, and Vernon and I got involved through our contacts at UNSW. And for much less money, we produced something much more effective for them.[45] |
| Chris Starmer | Charlie [Plott]. |

[43] The project was conducted for the Federal Government of Australia in the period 1996–2006 and was titled "Testing a proposed system scenario for negotiating energy delivery contracts in real time."

[44] There is no publicly available report by the *National Grid Management Council.*

[45] **Chao, Hung-po and Hillard G. Huntington.** 1998. *Designing Competitive Electricity Markets.* Dordrecht: Kluwer Academic Publishers, **Denton, Michael J.; Stephen J. Rassenti; Vernon L. Smith and Steven R. Backerman.** 2001. "Market Power in a Deregulated Electrical Industry." *Decision Support Systems*, 30(3), 357–81, **Olson, Mark; Stephen Rassenti; Mary Rigdon and Vernon Smith.** 2003. "Market Design and Human Trading Behavior in Electricity Markets." *IIE Transactions*, 35(9), 833–49, **Rassenti, S. J. and V. L. Smith.** 1998. "Deregulating Electric Power: Market Design Issues and Experiments." *International Series in Operations Research and Management Science*, (13), 105–20, **Rassenti, Stephen J.; Vernon L. Smith and Bart J. Wilson.** 2003. "Controlling Market Power and Price Spikes in Electricity Networks: Demand-Side Bidding." *Proceedings of the National Academy of Sciences of the United States of America*, 100(5), 2998–3003, ____. 2002. "Using Experiments to Inform the Privatization/Deregulation Movement in Electricity." *CATO Journal*, 21, 515–44.

## Contract Research: Putting Careers at Risk

Charlie Plott — I can address part of this question in terms of how often do you turn these things down? I frequently see things I won't touch because I don't think I could solve the problem. And if you take one of these things on, and you fail, you just cost somebody his career. So going in this direction is not lightly taken. But in terms of the methodology [of role playing or simulations of very complex processes], it is interesting because there is something in the military called gaming, which is essentially this, and they spend millions and millions and millions of dollars on gaming. [Hoffman: Billions]

And *Martin Shubik* was taking this out of in a big paper on how the military is wasting money on gaming.[46] And we have been brushed with an unfavorable brush stroke by this kind of literature because of the reputation of these people doing this kind of research, and they get paid a lot of money for it.

Stephen Rassenti — Yeah. I can attest to that, too. Recently we had a contract from the Navy to look at rescheduling of personnel in the Navy.[47] And we were there competing with a lot of agencies that do large scale agent based simulations. [Plott: yeah] And those guys are getting three or four times the amount of money we were asking for. And right now, we are out of the money, and they are in.

Chris Starmer — And the situation you are describing of having a good flow of people coming to you with projects, is that a relatively recent thing? Is that something you had to develop? What is the history of this? When did you start doing these things?

Stephen Rassenti — We started doing these in the power industry. I think the first one we did was for the state of Arizona, right Vernon?[LVI]

Vernon Smith — Yes. And that was in '84.

Betsy Hoffman — Yes that was when I was working there.

Vernon Smith — In 1984, a one year contract. Really, that is what got us started on electric power decentralization issues.

Stephen Rassenti — Right. And that has persisted. We got a continuous flow dribbling in once every two or three years over the last 15 years in the electric power industry. There are still an amazing number of configurations of real world markets for the distribution of electric power, and they haven't converged anywhere to a uniform way of doing this thing. For example, in Singapore,

---

[46] A precise reference has not been determined.

[47] Project title: Examining Navy personnel reassignment practices. U.S. Navy January 2007—December 2009.

| | |
|---|---|
| | they still have vertically integrated units that have to create a Chinese wall, and the production part sells to the system, and the distribution part buys from the same system, yet they talk to one another on a daily basis. In the U.S., we have a deregulated production site, but on a regulated distribution element. There are all kinds of things to model in the laboratory. And clearly, no one has converged on the way this should be done. |
| Chris Starmer | Betsy, you wanted to comment. |
| Betsy Hoffman | Well, had I not gotten into administration and stayed with Vernon's group, I probably would be continuing to do this. But I sort of have a counter example. I got involved with a law firm that asked a group of us to do a series of experiments related to willingness to pay, willingness to accept contingent valuation. And we did a lot of experiments.[LVII] |
| Chris Starmer | When was this? |
| Betsy Hoffman | This was in the late 1980s. It was just before I went over to the dark side as Vernon likes to say. It left a really bad taste in my mouth because I couldn't publish the results. Then when the results were really good, what really galled me was we were hired by *Exxon* in the Exxon Valdez case[48] to basically do a set of experiments to show that contingent valuation was not a reliable way to estimate valuation.[49] And we got extraordinarily good data. They wouldn't let us publish it. The next thing I knew, they had given it to [*Peter*] *Diamond* and a couple of other people. They were going to publish our data under *Diamond's* name. He had done the econometric analysis.<br>We had done econometric analysis as well, and they were going to let him publish it under his name. And I approached him at an *American Economic Association* meeting and I said, "Do you know where this data came from?" He said, "No, Exxon gave it to me." I said, "I ran those experiments. I designed these experiments. This is my data." |
| Chris Starmer | And just to clarify, when you say you got very good data, by that – |
| Betsy Hoffman | The data we got very clearly showed that what people said they were willing to pay was different from what they actually were willing to pay by a factor of about 10. In other words, their actual willingness to pay was about one-tenth their real [stated] willingness to pay. And they were very carefully designed experiments. [Smith: I remember that] I think one |

[48] The Exxon Valdez oil spill was one of the worst environmental disasters of the 20th century. It occurred in Prince William Sound, Alaska, on March 24, 1989.

[49] Exxon hired Betsy Hoffman in 1991.

of the things that I have always prided myself on is being a fanatic about experimental design and utter and complete, sometimes to Vernon's distraction, fanatic about experimental design. These were very, very well designed experiments. And the data was outstanding data.

Because Exxon owned the data, they were literally going to give it to *Diamond*, and I think it was a couple of other people, and let them publish once it had been entered into the expert witness file for Exxon. Of course, I was very happy to have Exxon lose big after all of this. They were going to give it to them and let them publish them. I approached him at a meeting and said, "Do you know where this data came from?" And they said, "No." I presented them with our paper, and I said, "This is our data."[50] To *Peter Diamond's* credit, he was horrified. He was utterly horrified. Economists are actually very respectful of one another's data. I actually have two publications[51] now out of that data but only because Peter told Exxon that –

Vernon Smith: That this was immoral.

Betsy Hoffman: Right.

Vernon Smith: Very simple.

Betsy Hoffman: Right. That he would not publish his work, his econometric analysis of our data if we were not allowed to publish the experimental results. After that, I was fed up. I was fed up with doing contract research. It wasn't worth it.

Chris Starmer: Charlie [Holt].

## Continuities Between the Lab and the World

Charlie Holt: In my experience, some with John Ledyard and *Jacob Goeree* has been like with the FCC. They did let us publish results based on our contract work.[52] For the *Regional Greenhouse*

[50] At that time it was a working paper which became: **Binger, Brian R.; R. F. Copple and Elizabeth Hoffman.** 1995a. "Contingent Valuation Methodology in the Natural Resource Damage Regulatory Proces: Choice Theory and the Embedding Phenomenon." *Natural Resources Journal,* 35(3), 443–59.

[51] See Endnote #LVII.

[52] **Goeree, Jacob; Charles A. Holt and John O. Ledyard.** 2007. "An Experimental Comparison of Flexible and Tiered Package Bidding," Wireless Telecommunications Bureau of the Federal Communications Commission. ____. 2006. "An Experimental Comparison of the FCC's Combinatorial and Non-Combinatorial Simultaneous Multiple Round Auctions," Wireless Telecommunications Bureau of the Federal Communications Commission. ____. 2004. "An Experimental Investigation of the Threshold Problem with Hierarchical Package Bidding," Wireless Telecommunications Bureau of the Federal Communications Commission.

*Gas Initiative* where we designed auctions for them, it was a contract through the University of Virginia, and the university negotiated with a lot of force on our behalf. And they came up with an agreement that we could publish anything we wanted from the data one year after the first auction. See, they had the idea that people could look at this data and figure out how to bid in an auction. [Ledyard: right] [some laughter] So that would be irrelevant a year after the first auction. But I thought that worked very well.[LVIII] [Rassenti: yeah]

The other contract I worked on recently with Charlie Plott and *Jacob Goeree* was a lot of fun, it lasted only about a week and a half. We got a contract to design the TARP auction for the treasury.[53] And that was just a blast. One Monday morning they called and said, "Why don't you guys just take a vacation." They couldn't tell us it [the auction design] was suspended because the stock market was too volatile. But ever since, it has been just an academic project.[54] We have been funding it ourselves, doing our own experiments.

What I love about contract research is you come into contact with a real problem. You get an inside view of what the real issues are. And for a long time, I would do experiments in the lab. I was just perfectly happy in a laboratory to try to understand how people behaved in auctions, games and the markets. If it ever applied to the real world, I was thrilled. And then finally, people started asking questions, what if we ran this type of an auction? They will give it a name. They will give it an acronym. They will start calling you and asking you how you program it. When people on the outside world are interested in implementation, that is a whole new dimension of reality, which I think is a great source of discipline for experimenters.[55]

Chris Starmer [Addressing Charlie Plott] This is something you have been waiting to come in on for some time.

Charlie Plott I was just going to make a comment on that. The thing that I have discovered, when you build this up from the labs, and then actually implement it, is that it is extraordinary how your

[53] TARP (Troubled Assets Relief Program) is a U.S. government program established under the *Emergency Economic Stabilization Act* of 2008 with the specific goal of stabilizing the United States financial system and preventing a systemic collapse by purchasing assets and equity from financial institutions.

[54] **Armantier, Olivier; Charles A. Holt and Charles R. Plott.** 2010. "A Reverse Auction for Toxic Assets." *Social Science Working Paper, California Institute of Technology*, 1330.

[55] Pertains to the Goeree and Holt (2010) Hierarchical Package Bidding paper cited in Chap. 3, Footnote 76.

learning from the lab applies to the real world. It is just like a little bigger experiment. That is all. And [the real world] behaves just like a little bigger experiment. The continuity is phenomenal. But on the contract research, of course, I have done a lot of that, and I haven't had a bad experience yet.

Chris Starmer: Vernon.

Vernon Smith: Well, just to add to what Charlie [Plott] said. You get these practitioners who really know their business well. And the laboratory provides you an opportunity to think together about how you design the experiment and how you do it. One of the things you don't get from these guys is that there is something irrelevant about these experiments. You don't get that from them. [Plott: That is true] They take it seriously. If they are worried about subjects, they get their own and put their own people in it—that sort of thing.[LIX]

## The (Missing) Boilerplate in Contracts

On the thing that Betsy mentioned with *Exxon*. What we do is to tell the client upfront that we have to publish this. In a university environment we can't be doing this research.

Betsy Hoffman: I did that, too.

Vernon Smith: With the data? [Hoffman: Ehm] Well, they just screwed you over. They are typically reluctant. They want to have it proprietary. I recall one power company, we got them on board by saying look, we are going to disguise anything that is your data. We are going to disguise all of this; we can scale it. There are all kinds of things we can do. And also, you can read the paper and make sure you are comfortable with it. They were not only comfortable with it. In the end, they wanted credit for having funded it.[56] [some laughter]

Stephen Rassenti: In the publication.

Vernon Smith: Yeah. And it is in the publication. It is important to deal with this upfront. [Rassenti: ehm] We had to turn down stuff, really big stuff, because they wanted it strictly proprietary. Don't go there. You have got to realize that it is not research, it is consulting. They are going to keep it, and I don't think it is satisfying.

---

[56] **Editor**: "What was the company and what was the publication? **Vernon Smith**: "This was Ohio Edison in the late '90s. I don't recall which of our papers on electricity involved that study, but they were very involved in that study and liked it. Of course by then we had been to New Zealand and Australia and knew the ropes."

Chris Starmer: Betsy, two hands.

Betsy Hoffman: Well, I think what happened to me was I wasn't considered famous enough back then. *Peter Diamond* [was famous enough]. The results were so good that they didn't want somebody that they didn't think was famous enough to have their name on this paper that was going to be their central piece of their legal argument. They wanted a famous name on it. I had an agreement that I could publish it. But without telling me they were basically going to let somebody else publish my data before I published it, and, therefore, render my data useless. I had to intervene with the person himself. Essentially make him realize that what he was being asked to do was immoral.

It wasn't that I wasn't going to be allowed to publish, it was that they were going to let somebody else publish it before me, which meant that it was no longer considered to be my data.

Chris Starmer: Al, You have been waiting some time.

Al Roth: I agree it is important with contracts to make it clear that professors can't sign contracts that won't let them publish. Sometimes the boilerplate in contracts I have seen says much more than that. They say you are going to work for us on market design, and you will learn all sorts of secret stuff from us. And so you won't thereafter talk or write to anyone about market design. And you say no, no, no. That would put me out of business. On the other hand, often you can have standard clauses of your own, which say university professor, we preserve the confidentiality of your data, but we are going to publish what we need to.

What I was going to say about applied work and experiments is that in doing market design that goes into operation, there are lots of people who have to be persuaded of lots of things. What I found is that experiments sometimes play a less heroic role than I might once have imagined they do. That they work well in concert with lots of other things. The experiments will be a piece of the winning argument but seldom the whole winning argument. As experimentation has matured and as market design is starting to come online, we see this when we talk to school board administrators when we are designing municipal school choice.[57]

[57] The reference for the medical labor market is: **Roth, Alvin E. and Elliott Peranson.** 1999. "The Redesign of the Matching Market for American Physicians: Some Engineering Aspects of Economic Design." *American Economic Review*, 89(4), 748–80. The references for school choice mechanisms: **Abdulkadiroglu, Atila; Parag A. Pathak and Alvin E. Roth.** 2005a. "The New York City High School Match." Ibid.95(2), 364–67, **Abdulkadiroglu, Atila; Parag**

We see this when we talk to professional organizations. When we are designing labor market clearing houses. Experiments are definitely playing a useful role, but it is a modest one. You have to be careful not to over claim what the experiment accomplishes in that.

## The Experiment as Interface for Arguments

Stephen Rassenti Considering what Al just said, that goes even further. For example, in the power industry, I find that a lot of times what is most important is that the people who hired you to do this actually get to participate in it, think about the problem, and play as agents in the system that you are designing. Then think about what they really want to do out there in the real world. Sometimes that is much more important than the results you generate for them. The fact that they can use the system that you design and have various people that are affected by the decision making process participate also in the development and the experiments themselves.

Chris Starmer Charlie, on the way in here this morning, if I got the spirit of what you were saying right, and correct me if not, I took you to be saying that the applied research with experiments has had a really important impact in just demonstrating the importance of the experimental method.

Charlie Plott Yeah. Somebody might disagree with me, but I think that the applied work heavily contributed to the success and the development of laboratory experimental methods in the '80s. This early work basically had the function of showing the profession that it was good for something. It was one after another, after another, sometimes big, sometimes small, and sometimes quite visible like the airport study. It basically demonstrated to the profession that this was more than handing out money to kids. That it had results that actually had an impact and was useful.

---

**A. Pathak; Alvin E. Roth and Tayfun Sönmez.** 2005b. "The Boston Public School Match." *American Economic Review. Papers and Proceedings*, 95(2), 368–71.

A chapter on experiments in market design is included in the forthcoming *The Handbook of Experimental Economics*, Volume 2, with John H. Kagel and Alvin E. Roth as editors.

That message was certainly in the paper I published in the *Journal of Economic Literature*, which had some impact.[58] Actually it was commissioned because I did a similar paper for the *Federal Trade Commission*.[59] The *Journal of Economic Literature* saw that and said, "Can you do something like that for industrial organization?" Part of that [paper] was a story that you have this very basic research that you might have thought just had no use other than for scientists. But look how this curiosity driven research all of a sudden springs in unexpected ways into valuable things. That was the message out of that paper and whole series of these things. And I think it was really important. That was one of the things that launched experimental economics.

Betsy Hoffman: I would agree.

Vernon Smith: Funding aids have really changed dramatically when we went electronic. Up to that time, you would get some money for subjects, maybe for summer salary or something like that. It was fairly modest. We started doing computerized experiments in 1976 at Arizona. Fortunately, the university was a PLATO site, and so we were able to use that. We used funding from the university.

Chris Starmer: I think talking about PLATO, the technological side of this and perhaps the funding of it is one of the things that I would like to cover tomorrow. Rather than getting into that at the late stage of today, because I think there are a number of interesting things to explore there, I would prefer not to compress into a very small space of time. We are reaching the end of our allotted time for today, [and the organizers are] nodding at me from over in the corner. I will bring proceedings to a close for today. Thank you all very much for your contributions and being such a cordial group to be chairing.

---

[58] **Plott, Charles R.** 1982. "Industrial Organization Theory and Experimental Economics." *Journal of Economic Literature*, 20(4), 1485–527.

[59] **____.** 1981. "Theories of Industrial Organization as Explanations of Experimental Market Behavior," S. C. Salop, *Strategy, Predation, and Antitrust Analysis.* Federal Trade Commission.

# Chapter 5
# Knowledge and Skills

Chris Starmer Good morning, everyone. Welcome back. Congratulations to us for assembling a complete group by more or less 8:30 in the morning after a late, but extremely enjoyable dinner. One thought that crosses my mind is that I am going to try to encourage you where I can in the sessions today, as I was doing yesterday, but perhaps doing more so, to illustrate answers that you are giving me, points that you are making with reference to specific examples of your work, other people's work, when these things were, and what happened. I want to encourage you to think in that reminiscence mode and provide examples that you are drawing on.

The organizers would particularly welcome answers, illustrations that perhaps draw on things that may be relevant and interesting, but you think may not be part of any formal record as yet and may otherwise never be part of a record. In terms of the specific aims of today's sessions, there are three sessions. We are considering, perhaps, adjusting the length of the second session from one and a half hours to an hour and giving us potential for a longer third session. We may do that. But we have three sessions covering two topics today.

The first is broadly speaking related to skills, the second related to issues with labs. In relation to skills, knowledge, what our skills are, how we learn them, and how we pass them on, it seems to me that when we think about our skills as experimental economists, when we talk about them, some of the things we think of as skills are very obvious, come to mind easily, and are very apparent. For instance, quite often you need people with skills writing computer code. These are very



A. Svorenčík, H. Maas (eds.), *The Making of Experimental Economics*,
DOI 10.1007/978-3-319-20952-4_5

obvious, concrete things. Other times when we talk about skills of experimental economists, it seems to me that people speak of certain skills as being a bit elusive, a bit hard to quantify.

You know it when you have got it, but it is hard to say exactly what it is. I'm interested to explore both of those dimensions. As a way of getting into that, but without making a clear distinction between the obvious and the elusive, I would like to ask you to think about from your experience examples of experiments that you were associated with that you came to think, you formulated your own view, that in some sense they had gone wrong or failed. This was your verdict. Not the verdict, let's say, of some crazed referee. You thought it went wrong, and I want to add something else, that you think you learned something from it.

It led you, perhaps, overnight, the next day, perhaps some years later when you suddenly went back to your drawer to take that out again, that you drew something from that you think was significant. Perhaps if I could have volunteers. I'm happy to return to any examples at any point later in this session. You may think in three quarters of an hour, ah, I have got an example. Feel free to come back. But is there someone who would like to start us off with an example? Jim.

## Learning from Failures

Jim Friedman

This is an example that dates to roughly 1966. At the *Cowles Foundation* at the time was a very good collection of young theorists, one of whom was *Menahem Yaari*. He had, at that point, written a decision under uncertainty paper, which was a nice interesting piece. I don't exactly now remember the thrust of it.[1] But he and I talked about running an experiment to see if whatever behavior it was that was the content of the paper could be seen in the laboratory.

I recall [in this experiment] the subjects were looking at a collection of bets. Say you have a chance to win a dollar with a probability of 0.3, and you have a piece of paper in front of you that lists amounts of money from \$0.00, \$0.10, \$0.20 and so on cents down to some much too large number. And as a subject, you circle how much you would be willing to pay, just willing to pay to have that bet. You would get a sheet with a

---

[1] **Yaari, Menahem E.** 1965. "Convexity in the Theory of Choice under Risk." *The Quarterly Journal of Economics*, 79(2), 278–90.

large number of bets on them, and I don't remember a lot more than that. I'm sure knowing my predilection from my first experimental work, I know I would have wanted the subjects to be making decisions for a long time to get used to the experimental setting.

We were going to have very nice indifference curves coming out of what the subjects did. What we had was, as far as we could tell at the time, just a whole boatload of random trash. And that experiment never saw the light of day. We never figured out a way to redo it that we thought would be workable. I'm not sure what lesson I carried away from that beyond the fact that it was a failed, just utterly failed thing and a dead end.

Chris Starmer Did it steer you in ways to avoid doing similar things?

Jim Friedman It probably sensitized me more to the notion of having careful pilot experiments. But beyond that nothing that I can recall.

Chris Starmer Thank you. Betsy.

Betsy Hoffman Well, I'm going to report two experiments, one of which has never seen the light of day, and one of which actually did and became one of more cited papers. But then, because of criticism that Charlie made of it, I ended up doing another paper that led to a huge string of papers, including my most cited paper. The one that never saw the light of day was after Vernon and *Kevin* [*McCabe*] and I had done our double anonymous experiments, we decided that we were going to try computerizing the anonymous—remember that, Vernon?—the anonymous dictator experiments.[2]

We had these elaborate rules where every subject was given a random number generated by the computer that was their ID number. And we had the extra two envelopes with no money, with just white slips of paper. We replicated everything [Smith: Yeah] about the double anonymous dictator experiment,[3] and we got utter trash for results. Basically what we got were results that were undistinguishable from dictator results in which the experimenter can observe everything. I don't know, did you and *Kevin* go on to ever explore that and publish anything out of it?

Vernon Smith No. And I think what was interesting was how replicable the results were when people could see the way it was, in a sense,

[2] **Hoffman, Elizabeth; Kevin McCabe; Keith Shachat and Vernon L. Smith.** 1994. "Preferences, Property Rights, and Anonymity in Bargaining Games." *Games and Economic Behavior*, 7 (3), 346–80.

[3] In an *ultimatum game*, player 1 makes an offer of $X from a total of $M to player 2. If player 2 accepts the offer, then player 1 is paid $(M-X) and player 2 receives $X; if player 2 rejects the offer, each gets zero. In the *dictator game* the offer made by player 1 is final.

acted out when you ran it by hand. It was obvious that no one could know.[LX]

Betsy Hoffman: I think that our hypothesis was, and this might be something to explore, and actually, *Andrej* [*Svorenčík*] asked me if we had ever published this paper because I mentioned it. Our hypothesis was that basically, subjects do not believe that anything that is on a computer is truly anonymous, even with the random numbers, even with the sealed envelopes, even with the 20 white slips of paper. Everything else replicated, we were not able to replicate the result that people took all the money. We never explored that further to figure out what happened.

Vernon Smith: Well, and it is a good example of the unpublishability of stuff you learn the most about.

Betsy Hoffman: Yeah. Right. Exactly.

Vernon Smith: Because people expect a result. We just –

Betsy Hoffman: Well, we got a result.

Vernon Smith: Yeah, we got a result. But that is not what, you see –

Betsy Hoffman: So then the other one is the Coase paper,[4] which is actually among my top five cited papers. But the result that we got on the Coase paper was that people split the profits equally. And Charlie [Plott] calls me up one day and says that paper you just published in the *Journal of Law and Economics* that is getting so much play, he said, "you didn't test the Coase theorem, because the Coase theorem [states that] the subjects are actually supposed to divide it according to the core." And it wasn't divided according to the core, it was an equal split. So Charlie says to me: "I'm going to write a nasty comment." [Addressing Plott] I don't know if you remember this or not –

Charlie Plott: Yeah. I remember this.

Betsy Hoffman: He said, "I'm going to write a nasty comment and send it to the *Journal of Law and Economics* about why your results are trash." And I said, "just give us a chance to explore this." *Matt* [*Spitzer*] and I put our heads together and came up with the game that where you had to win the right to be the controller. Then we also tested it as you had to win the right and then be reinforced by being told that you had won the right. By that we designed a whole new experimental technique, which tens of other people have used to elicit a belief that you have the right to behave in an individually rational fashion and were able to show it.

Then we reran those experiments, and we got the core result in the environment in which people had to earn the

[4] **Hoffman, Elizabeth and Matthew L. Spitzer.** 1982. "The Coase Theorem: Some Experimental Tests." *Journal of Law and Economics*, 25(1), 73–98.

| | |
|---|---|
| | right and we reinforced it by telling them that they had earned the right. Then that became *Entitlements, Rights, and Fairness* which is again among the top five cited papers leading to *Preferences, Property Rights, and Anonymity* which is my most cited paper.[5] |
| Chris Starmer | Thank you for the nice example. Charlie, did you have a hand up moments ago? |
| Charlie Plott | No. I have never had a problem with an experiment. [group laughter] |
| Betsy Hoffman | Charlie has never had a problem with a referee either. |
| John Ledyard | That is true. |
| Vernon Smith | Well, that is not quite right, Charlie. [group laughter] I will tell the way it is. [group laughter] [Plott: Okay] [You were] absolutely convinced that you couldn't get bubbles. [Hoffman: Yeah] And so *Dave Porter* went off and told Charlie I want to come over and I will show you how you will get a bubble. He ran [it in] Charlie's class, and Charlie has this idea to make them write down, every period what the next period's value is, because there is fewer dividends [to use] up. And, of course there were bubbles. [Plott: Big bubbles] And Charlie looks out there. Dave goes out there [to check what subjects are doing behind their computers]. They are not writing it down. So Dave goes out and says you are supposed to write that down. [Smith hits the desk repeatedly with his finger] The guy says: I'm trading, [Smith slaps the paper notebook on the desk] you write it down. [implying a maniac behavior, prolonged group laughter]. |
| Charlie Holt | That is a good one. |
| Charlie Plott | I think in terms of problems and methods, I don't know whether I can give any generalizations. Of course, I have had hundreds of bad experiments. |
| Vernon Smith | Can you speak up? |
| Charlie Plott | [bit louder] That was one of my [interrupted] |
| John Ledyard | It's on the record. [group laughter] |
| Vernon Smith: | Why are you whispering? |
| John Ledyard | It's on the record now. We've heard him. |
| Betsy Hoffman | Don't stop him. Keep him going! Come on! |
| Charlie Plott | In most of them I learned that if the model is not working after the first round or two, go check to see what is going on in your experiments. And very frequently, you are going to find somebody who is systematically confused. |
| Chris Starmer | Can you give me an example of where this happened? |

[5] ____. 1985. "Entitlements, Rights, and Fairness: An Experimental Examination of Subjects' Concepts of Distributive Justice." *The Journal of Legal Studies*, 14(2).The other paper is referenced in the Footnote 4 in this chapter.

## Confused Subjects and the Logistics of an Experiment

Charlie Plott: Well, it is easy. I remember once when we first started doing computerized experiments, it is just a regular demand and supply.

Chris Starmer: When would that be?

Charlie Plott: When would that be?

Chris Starmer: Yeah.

Charlie Plott: That would have been late ‘80s. And it wasn’t converging because we had one subject who was a supplier and she wasn’t selling anything at all. She was sitting there. She was [a student from] Russia and, of course, the market was reacting. We knew exactly what was happening because in supply and demand, if one of those suppliers is not supplying, it goes to a different equilibrium. The equilibration typically will tell you where something is going on that is non-parameter based. The girl was Russian, and my research assistant happened to be Russian as well. He went over and chatted with her. And he just said [to her] in Russian, “You should watch the screen.”

She should have been watching the screen because the quotes were coming across. With that instruction, she immediately [Plott snaps his fingers] turned around and the market price went up—it started to work just right. Now, you could add similar examples to that one. Another is the case in which subjects are given an [incentive] chart[6] where their payoffs go down as units go up, but they will [read] across the page. Basically, then you are running an experiment that has completely different parameters than you thought. If you have an experiment where the subjects are buying from the experimenter and selling to somebody else, some buy from the experimenter and forget to sell.[LXI]

There is a series of types of mistakes that can be made and I’m just talking about markets. Some errors are typical of people who have a misconception about their incentives or the types of information that might be available to them. If you are not careful giving instructions you have little idea of what the experiment is that you are conducting. When that is suspected we try to document the nature of the confusion and the source through the use of pilots. We then move towards a standardized set of procedures.

For example, if instructions go on the board, they go on the board from left to right, not spotted around, as sometimes a research assistant will do. You make sure that the writing on the board is large enough that it can be seen. You make sure that

[6] An instruction sheet with numerical values.

the [experimenter] reads from the board and watches the subjects because if the subjects are sitting out there doing something other than watching you, they are probably not listening to you. After the instructions, he should find out whether they have questions. If subjects don't have questions, chances are they don't understand. So there is a series of little skills and techniques that one begins to develop inside the laboratory to correct for what I would consider to be a loss of control in ways that result in the study of a phenomenon I don't care about studying.

Chris Starmer Could I ask other examples people can think of that have a more applied feel to them, perhaps, coming from consulting work. We haven't heard any examples of that.

Charlie Plott I don't understand the question.

Chris Starmer I'm going back to asking for examples of failed experiments and what you have learned from them, and whether there are examples that are related to more applied aspects of experimental research.

Charlie Plott Yeah. The first time we did pipelines, we were interested in a tâtonnement system. This was a network pipeline and we were just trying a tâtonnement system of nodes of a pipeline. And it turns out, of course, tâtonnement can have cheap talk. The system never converged because the cheap talk never got coordinated against nodes of the system. It was obvious that this kind of system was not going to work. We reported that that type of system was not going to work, and then we proceeded to design a different kind of system to carry through a pipeline system that would work.[7]

Combinatorial analysis. There are many of them where things don't stop. In these complex things there have to be stopping rules. Many, many, many times we discovered that the stopping facilities aren't appropriate. And the experiments will reveal the poor institutions. That is one class of things. Lack of control, that is a different thing where the subjects just don't understand. I would really separate those two out.

Chris Starmer Vernon, a moment ago, you said you have got many examples of how you can't publish the stuff you have learned most from –

Vernon Smith Yeah. The way I would say it is that sometimes it was harder to publish because where you are getting what is considered negative results, well, it turns out that it is giving you a lot of insight.

Chris Starmer Can you give me another example?

---

[7] **Plott, Charles R.** 1988. "Research on Pricing in a Gas Transportation Network," *Office of Economic Policy Technical Report no. 88-2*. Washington, D.C.: Federal Energy Regulatory Commission.

Vernon Smith: Well, when we first started doing—we had been doing supply and demand experiments with a lot of different institutions. And all of those experiments, it is a series of one-shot plays, and there is nothing that carries over, because you are talking about a market for hamburgers or haircuts. And [those items] cannot be re-traded whatever the institution is. You get the consequences, the realizations [of purchases]. People incur the costs and the values, and then it starts again. And you repeat. The first time we did those markets, they converged to a competitive equilibrium that subjects, of course, knew nothing about far faster than we expected. It was a case where, in a sense, it violated all kinds of beliefs and teaching at the time because people didn't expect it to work that well. That was in a way a contrary result.

Chris Starmer: When was this?

Vernon Smith: That first experiment was done in January 1956. After we got computerized, we started doing electronic trading. The first programming began in 1975. The first experiment is in 1976, and that was with *Arlington Williams*. And when Steve came on—[addressing Rassenti] you were about '78, '77/'78? [Rassenti: That is right] We were going electronic on virtually all the experiments. Not quite, but certainly in the market experiments.

We wanted to do some asset trading experiments. Now, here you have cash and shares, and their life is the entire life of the experiment. The values of the shares are determined by dividend drawings. Since it is a finite horizon, so you are using up draws in successive periods. Well, we started out—we wanted to find an environment so transparent that we would get trading at fundamental value. Then the idea was to see what we could do to create bubbles. Well, the first experiment we did in the transparent environment, we had huge deviation of fundamental value. And so we started to lean on them more, in a sense, give them more information.

Make them calculate so they could understand exactly what they were doing. No. It doesn't work. Charlie didn't believe these and started doing them. *Colin Camerer* says it is an Arizona phenomenon. So Colin does them at the University of Pennsylvania, and it turns out that at University of Pennsylvania it happens also. And it turns out that businessmen and all kinds of subjects– very robust on that—we are back to the subjects. In frustration, I recruited some over the counter traders in Chicago. This was 1989, and we got access to the computer up there. *Joe Swanson* got the subjects, so we went up and put traders in one of these experiments, and we got a massive bubble.

Moreover, it didn't collapse at the end because we thought these are sophisticated people. We will let them short sell and buy on margin. The problem with short selling is if your timing is not any good, all it does is just make everything worse. And this one doesn't collapse because everybody is buying to cover. It doesn't collapse at the end because the short position is high, and they are buying it every time it starts to go back to the equilibrium.

We really learned a lot about something that we did not begin to anticipate. These things are always harder to publish. After about the third round [of refereeing], we finally got *Econometrica* to accept it.[8, LXII] But there were still all kinds of people who thought there was something wrong with those experiments.

Chris Starmer: It seems clear from these examples that experiments generate many surprises and things that you really don't expect to happen when you go into them. [Plott: Oh yeah] Frans.

## Institutional Resistance to Experimental Results

Frans v. Winden: I have an experience that goes back to when we were commissioned to do this project on changing the tax system. It was based on a motion in Parliament that asked the *Minister of Social Affairs* [*and Employment*] to commission a project with CREED to investigate the performance of the so called van Elswijk Plan concerning a reform of the Dutch tax system.[9] Van Elswijk maintained the view that for social insurance purposes, it would be better from an economic point of view to change from a wage tax system to a value-added tax system, and to give the employer a subsidy for each employee as a bonus [for the saved unemployment benefit.]

We designed an experiment in consultation with a steering group that was made up by well known and academically

[8] **Smith, Vernon L.; Gerry L. Suchanek and Arlington W. Williams.** 1988. "Bubbles, Crashes, and Endogenous Expectations in Experimental Spot Asset Markets." *Econometrica*, 56(5).

[9] 1998/99: Dutch Ministry of Social Affairs and Employment: research project "An experimental study of the Van Elswijk Plan," concerning an alternative tax system for financing unemployment benefit Report 1999 http://www.prohef.nl/docs/summary%20creed.pdf [Accessed on March 31, 2015]. Piet van Elswijk was at the time affiliated with the Free University in Amsterdam (Economics Faculty) and the private consultancy firm IME Consult.

reputed experts in the fields of labor economics, public economics, and game theory. All the relevant skills were there. The other camp was made up by van Elswijk who had very specific ideas about what should come out of the experiment. We asked the two groups to come up with explicit hypotheses concerning the outcome of the experiment, and they were really different, countering each other. We also asked the steering group at every stage to agree with the design.

Then we ran the experiments and the results were very much in line with van Elswijk['s hypotheses], but not with the steering group's predictions, nor with the expectations of the Ministry. It became very uncomfortable, because I was actually approached at some point by the officials to change things in the report. And although it was contracted that the whole report would be part of a series, they decided not to do it. Nevertheless, we published several papers out of it as we had retained academic freedom of publication. There will be one still forthcoming in the *European Economic Review*.[10]

They started to quibble about all the results, despite they agreed on the design and were specifically asked about their hypotheses. Then they said, "Well, okay. But you should not take these hypotheses too personally. It means that we thought, as member of the steering group, what other people might think what the prediction would be." [They tried to separate] the link between what they said and the results that showed differently. It was actually quite an experience. We [CREED] said that, of course, given the fact alone that there were not many observations, our advice was not to implement the tax system, but that it would be worthwhile to further investigate it. But it was over. Then there was a letter written [from the Minister] to the Parliament reporting on the outcome of the study, where I couldn't find the results from our report attached to it.

I sent a letter to various Ministers such as the Minister of Finance and the chair of the Parliament saying that these are

[10] **Riedl, Arno and Frans van Winden.** 2001. "Does the Wage Tax System Cause Budget Deficits? A Macro-Economic Experiment." *Public Choice*, 109(3), 371–94, ____. 2007. "An Experimental Investigation of Wage Taxation and Unemployment in Closed and Open Economies." *European Economic Review*, 51(4), 871–900, ____. 2012. "Input Versus Output Taxation in an Experimental International Economy." *European Economic Review*, 56(2), 216–32, **van Winden, Frans; Arno Riedl; J. Wit and F. van Dijk.** 2001. "An Experimental Study of the Van Elswijk Plan: Value Added Taxation Instead of Wage Taxation as a Means to Finance Unemployment Benefits," Amsterdam: CREED, University of Amsterdam.

not our results, they are not in the report. But at that time, they were at an advanced stage of a planned tax reform, and there were already positions taken. We didn't hear anything about it anymore.

Chris Starmer — Okay. I have got three hands. Charlie [Plott], John, then Charlie [Holt].

Charlie Plott — Well, I'm not too sure what theme you want to push here. But there are certainly problems like where the customer in some sense didn't like the results and wanted to bury them. That happens.

Chris Starmer — Can you give me an example?

Charlie Plott — Sure. The barge problem[11] that we first did for the *Department of Transportation* was buried by the *Department of Transportation* because they were afraid they would end up with Proximire's Golden Fleece [Award][12] because it was so unusual and because it would become controversial. We published it anyway,[13] but the organization inside the *Department of Transportation* killed it.

## Isolating Confounding Factors and Learning from Them

I could give you many examples like that. There is another class though. You ask: "Are you surprised?" Well, I could give many examples where I was just flat surprised by the outcome. For example, the very first committee experiments we ran showing whether the committee under Robert's Rules would go to equilibrium, shocking. I didn't believe that. *Mo Fiorina* ran the first experiment. I thought that he had cheated, so I ran off at random, did it myself, and saw the same thing he did. A complete surprise.[14]

The rational expectations. The idea that you could run these markets, and with asymmetric information and the bids and asks would actually incorporate in the price the

[11] See Chapter 4, Footnote 16 for a reference and further context.

[12] U.S. Senator *William E. Proxmire* issued the Golden Fleece Award from 1975 until 1988. It identified what he considered wasteful government spending. NSF was one of the awardees and animal experiments by John Kagel and Ray Battalio was highlighted in his award report.

[13] The publication took place in 1982 with a delay caused by the refereeing process.

[14] See Chapter 2, Footnote 53 for a reference and further context on Fiorina and Plott (1978).

value of the asset, fully incorporate. We didn't believe that. Shocked when it happened.[15] Preference reversals. I didn't think we would see preference reversals when we actually controlled for everything. Boy, we sure did.[16]

Chris Starmer: I'm struck that sometimes what you are surprised by is the fact that the behavior is so different from what you expect in theory. I guess preference reversal would be an example of that. That just shouldn't happen. And other examples –[17]

Charlie Plott: It depends on who you are. These other examples, like rational expectations –

Chris Starmer: And quite the reverse.

Charlie Plott: Well, that is what the literature says is going to happen, and we didn't believe that.

Chris Starmer: Yeah.

Charlie Plott: [Take the] the equilibrium of majority rule. I developed that theory. All of that mathematics was mine. I didn't believe it. I thought Ledyard and other people were generalizing these to unbound spaces and continuums of people.[18] I just thought that was mental masturbation. I thought wouldn't it be fun if I showed my theory was wrong and showed their theory was wrong, too. Very destructive. Turns out my theory was right, and then I didn't know how to sell it because that is a little harder. In fact, that whole thing changed the way we approached the experimental method because I was a developer of a theory that now had a positive result.

If I reported it as a positive result, no one is going to believe me. How do I go about now understanding how to report a positive result? And so that led into a whole methodology that we finally used. Now, Vernon's bubbles experiment. That is really a good example, because I don't believe the bubbles. I

[15] **Forsythe, Robert; Thomas R. Palfrey and Charles R. Plott.** 1982. "Asset Valuation in an Experimental Market." *Econometrica*, 50(3), 537–67.

[16] **Grether, David M. and Charles R. Plott.** 1979. "Economic Theory of Choice and the Preference Reversal Phenomenon." *The American Economic Review*, 69(4), 623–38.

[17] **Editor:** Would be interesting to know the contrast that you wanted to raise. **Chris Starmer**: "I can't say with confidence this is what I had in mind, but here is my best shot. I have the hunch that a variety of things can provide benchmarks against which lab outcomes can seem 'surprising'. Theory provides one set of expectations—but sometimes you can be surprised by how well a theory does (not just failure). But less formal things provide benchmarks—informal conjectures, what you know about behavior outside the lab and so on. I think I was fishing for these sorts of things."

[18] **Hinich, Melvin J.; John O. Ledyard and Peter C. Ordeshook.** 1972. "Nonvoting and the Existence of Equilibrium under Majority Rule." *Journal of Economic Theory*, 4(2), 144–53, ____. 1973. "A Theory of Electoral Equilibrium: A Spatial Analysis Based on the Theory of Games." *The Journal of Politics*, 35(01).

didn't believe them. I still don't believe them. [some laughter] Even though I do them all the time, but there is an interesting story. I was watching Vernon's bubbles, and I said, "Let's do that to the opposite way."

I took Vernon's experiment. [Vernon: ehm] I made it isomorphic so that rather than paying a dividend, you had to pay a storage cost. Then Vernon's structure should burst up, you should see the same thing. [It] didn't happen. *Never* has happened. *Never* is going to happen.

Vernon Smith — We should get the mirror image of the bubble.

Charlie Plott — It should be the mirror image. I have never seen it. We should get it, but we don't. Then we started asking why. When you do the ordinary bubble experiment you get few questions but when you do instructions for the mirror image bubble, all of a sudden, these subjects just start asking all kinds of questions that show that they don't understand. I think that in Vernon's [bubble experiment], they are not really understanding what is going on here in a really deep, intuitive way. I said, "Let's test that. Why don't we run an experiment where you can bubble. But if you buy, you can't resell."[LXIII] So there is no speculating on the other person.[19]

Well, as it turns out, we can get bubbles. But you can't get them for the reason that Vernon and other people think because there is no resale here. It is not the sale of the other. [Smith: ehm] It is these guys are just flat confused about the dividend value. It is a very personal thing, not a speculation thing. That is the state of the art of why these things occur. I watch his [Vernon's] stuff with great interest, but I'm always on the sideline skeptical. And I also try to show that their theory is wrong and I think fundamentally that it is wrong, but I haven't been able to really prove it in good, solid ways.

But that is a very typical relationship you want to get a feeling for. It is a very healthy skepticism about replicating people's work and how the methodologies tend to evolve.

Vernon Smith — And it is important to do experiments to test your explanation of why it is going wrong, which is really what you are doing.

Charlie Plott: — Yeah.

Vernon Smith — You see, it didn't stop the bubbles. What it did was change the way you thought about why it might be coming back.

Chris Starmer — John.

[19] **Lei, Vivian; Charles N. Noussair and Charles R. Plott.** 2001. "Nonspeculative Bubbles in Experimental Asset Markets: Lack of Common Knowledge of Rationality Vs. Actual Irrationality." *Econometrica*, 69(4), 831–59.

## Costly Consequences of Actions: Who Will Pay for the Soufflés?

John Kagel — Well, when we did our first winner's curse experiments, there had been previous reports of winner's curse, and I think the first paper was by Bazerman and Samuelson.[20] They were looking at coins in a jar and this sort of thing. And there could be errors in the personal estimates of what the values were. We said, "Okay, look, what we are going to do is to give them a signal with the properties that the theory says it should have." Furthermore, we will provide them with complete feedback after each auction. We are going to list all the signals and all the bids and the true value, and we will have a couple of dry runs.

When *Dan* [*Levin*] and I started out, we thought that the winner's curse would go away after a couple of rounds. And of course, it doesn't go away.[LXIV] Then the first paper in that series was published after the second paper in the series because it had so much trouble going around.[21] There were issues raised like—was limited liability for losses responsible for the winner's curse? Charlie [Holt] tried something with eliminating the limited liability or attenuating it, and he replicated our results.[22, LXV]

Chris Starmer — And were these questions coming through referees' reports or seminars?

John Kagel — It came through referee reports and then a comment on—so the first paper couldn't get published. I mean, we had trouble getting it published. Then we said, "Okay, we will go back, and we will do experienced subjects." Let me give you a little anecdote here though because we were doing this at the University of Houston, and we were –

Chris Starmer — When was that?

John Kagel — This would be in '84, '83/'84. We are doing this at the University of Houston with MBAs. And one woman comes in to our experiment, and she says: "Is this the winner's curse experiment?" [group laughter] She went bankrupt in the sixth period or so. [group laughter]

---

[20] **Bazerman, Max H. and William F. Samuelson.** 1983. "I Won the Auction but Don't Want the Prize." *Journal of Conflict Resolution*, 27(4), 618–34.

[21] **Kagel, John H. and Dan Levin.** 1985. "Individual Bidder Behavior in First-Price Private Value Auctions." *Economics Letters*, 19(2), 125–28, ____. 1986. "The Winner's Curse and Public Information in Common Value Auctions." *The American Economic Review*, 76(5), 894–920. See also footnotes 153 and 155.

[22] **Holt, Charles A. and Roger Sherman.** 1994. "The Loser's Curse." Ibid.84(3), 642–52.

| | |
|---|---|
| Chris Starmer | You saw to that. |
| Betsy Hoffman | So she knew the theory and it wouldn't do her any good. |
| John Kagel | She couldn't apply it. But then the real puzzle for us, at least for me, was that we brought back experienced subjects. In small groups, they had overcome the winner's curse. They weren't anywhere near the Nash equilibrium, but then take the same subjects and you throw them into a larger group. They just go and, in fact, move in the wrong direction and commit winner's curse all over again. You start to have to think about how can this be? How can they get it in the one case but not in the other? And then it just crystallizes. This has to be context specific learning. They have not absorbed the theory at all. And that was a big insight.[LXVI] That was just a tremendous insight because it really shows up in a lot of different areas.<br>The other outgrowth of that was that the theorist didn't like the result at all. There was this idea that public information will raise revenue. And we showed that if there is a winner's curse, so you haven't satisfied the initial conditions for public information to raise revenue because they are losing their shirts to begin with, that it will, in fact, reduce revenue.<br>Over a couple of years, the theorists are now start writing things like well, if there is no winner's curse, it will raise revenue. If there is a winner's curse, [public information] will prevent people from going bankrupt, this sort of thing. Public information. So no attribution to our results, but I think they had an impact. |
| Chris Starmer | John, you are nodding away enthusiastically. Do you want to add anything to that? |
| John Ledyard | No, I'm going to stay away from that one. [Plott and Ledyard laugh] |
| Chris Starmer | Charlie [Holt]. |
| Charlie Holt | Well, this is a story related to some consulting. I was working with *Ron Cummings* at Georgia State and *Susan Laury*. Ron is the kind of guy you go to a restaurant, and they say you have to order soufflés 45 minutes in advance. He wanted to ask people at the table [unifished], he is so generous. Anyway, we are in the middle of this auction. We were designing an auction for the Georgia *Department of Environmental Protection*, and it is a reverse auction where the government is going to buy back irrigation permits from the farmers for the growing season because there is a big draught. We were helping them design the auction, and we were running experiments.<br>The idea is the government has a fixed budget. They will take the bids and select the low bids until they exhaust the |

budget. And so this was something where you could draw the demand and supply curves and the predicted price was about $110.00 per acre or something. It is a multi round auction, and part of the way through, one of the subjects says what happens if there is a tie at the cutoff. You use up all your budget, you take all these bids, but two people are tied.

What are you going to do? Well, Ron is the generous guy who orders all the soufflés, and says: well, if it is a tie, we will just include everybody. We will include the tie. [repeated oops by multiple people—Smith, Hoffman, and other, laughter by Ledyard] So there was a tie at 115 in one round. The next round, somebody else had a bigger tie at 115. Next one is even bigger. And the stack was getting bigger. [Smith: they were coordinating, weren't they?] We were spending twice the budget. Imagine you had a $10 million budget and you would spend $20 million. Ron said, "Let's change the rules." We said, "Oh no, you can't change the rules in the middle of the auction." That would mess up the subject pool, so we played it on and on and spent twice the budget.

But then, of course, that night, we did another one [experiment] where we selected randomly if there was a tie, and that came into the competitive price. We learned a lot from that, but those kinds of things help you write the rules of an auction. It is like computer code. If you mess up something small, it might all come out totally messed up.

Something similar happened when I was working for a government agency on spectrum auctions. They didn't want the database to record the old bids. And for this particular type of auction, if you didn't record the old bids, you might get a price cycle.[23] And so we just ran one in class and got a price cycle, and that was enough of an argument to go ahead and change the programming.

Chris Starmer: Vernon.

Vernon Smith: A comment on the question that has come up, do the subjects understand? Very often, it turns out what they don't understand is what *you* understand about the model and the situation.

Chris Starmer: Can you give me a concrete example?

Vernon Smith: Well, in the bubbles. You are supposed to understand that if this is the fundamental value, other people should not be

[23] This was published in the Goeree and Holt (2010) paper on the "paper and pencil" auction cited in Chap. 3, Footnote 76. The paper with Ron Cummings and Susan Laury was also published: **Cummings, R.; Susan Laury and Charles A. Holt.** 2004. "Using Laboratory Experiments for Policy Making: An Example from the Georgia Irrigation Reduction Auction." *Journal of Policy Analysis and Management*, 3(2), 341–63.

motivated to sell it at any different price. But you get these [deviant trades], and it is a mystery why it is that people get caught up in self-sustaining expectations of rising prices. In the housing bubble, we had all kinds of people were talking about the housing bubble. I remember one of the journalists said, "It can't be a housing bubble because everybody is talking about it."

No. *Arlington Williams* does these bubbles where there is a crossing twice a week, and he does them with his class.[LXVII] Okay. They talk about it. They come in: "Professor *Williams*, when is this going to break?" He says: "I don't know. You are doing it." And then they go back and the next cross is higher. So people are even challenging their own beliefs and things like that, you see. Well, the basic thing is not understood in terms of why this happens. One of the things you can show is more money, bigger bubble. You can model that. The momentum model predicts that if people have more cash, or they can buy on margin. These things give you bigger bubbles.

The important lesson in bubbles in the world is not whether you can stop them, but whether you can confine them to the people who are doing it. And we have solved that problem in stock markets. Basically, margin requirements, when you do have a major collapse of the stock markets, it doesn't devastate the economy. You have a collapse in the mortgage market, it devastates the economy because of the weight. There is some understanding here about that comes from that.[LXVIII]

## Anything You Change Can Make a Difference

Chris Starmer: I think this is an important point because it is probably taking us away [Smith: Yeah] from the issues of the skills in the context of experimentation. We have got about ten minutes left for this session. Perhaps if I just try with another quick theme to close this session. It seems to me that the different experimenters' practices differ to the extent to which they are close to the process of designing, implementing experiments, and running them. Some people write their own code. Some people perhaps run their own sessions or have other people to run their sessions.

I'm just interested in reflections about how close you have tended to be to different aspects of designing and running experiments and whether that has changed over your history of experimentation. Charlie.

Charlie Plott Yeah. With my students, I always participate in all their early experiments. I never let them run an experiment without me. And that typically has continuity across the group, so everyone has essentially the same thing. Then after I'm convinced that they know how to run an experiment, then I will start letting them do it themselves. But I'm always there personally.

Chris Starmer So then you let them loose, and they are coming to you with the idea, and they are implementing it, and then coming back with the results.

Charlie Plott Well, that is not the way it is. Most of the time these are teaching classes. The students have no idea what a good experiment is. In my class, I will tend to probe them and ask them to suggest [an experiment]. The things they suggest are like how am I going to solve world war, or how am I going to solve some big problem. But I try to get a feeling for what it is they are interested in. Then I pare that down, and pare that down, and pare that down, and get it smaller, and smaller, and smaller. Only when they finally have a little bitty hair of the problem they originally [found attractive], I say: "Okay, let's try to do that." And then we start designing [the experiment].

Now, they always know where they ultimately want to take that simple little thing. But then you start with something really, really, really simpler and even simpler, and then I'm stuck with them until I'm actually convinced that they know how to do the experiment. Then after that, then they are self sustaining. That [training] takes a typical student about a quarter. After a quarter I tend to trust them a little bit. It is a very intense training, like an apprenticeship essentially.

Chris Starmer Betsy.

Betsy Hoffman: I have always taken the approach that anything that you change can make a difference. If you change experimenters, if you change location, and if you are going to—so if I am going to do—trying to replicate an experiment or do a variation on an experiment. Since I have moved a lot, I have had a lot of experience with this. My view has always been that you have to be sure that your new subject pool and your new collaborators can get the same results as the previous subject pool and the previous collaborators. I always had a set of experiments that I would insist on replicating at the new place.

Chris Starmer Can you give me an example of where you have done this?

Betsy Hoffman With the ultimatum and dictator games, for example. The simple ultimatum game where there is no earning the right [to be the proposer]. You are randomly assigned a position of the proposer or the counterpart. There is a certain set of instructions. Before I start a new set of experiments on the theme of the ultimatum and dictator, I want to be sure that in the new context, I can replicate the original results of dividing the money equally.

Chris Starmer But perhaps I misunderstood, but I thought you said that when you moved to a new institution working with new people, you wanted to try and recreate the –

Betsy Hoffman Right. I want to make sure that the new environment replicates the results of the old environment.

Chris Starmer And I was interested in where you have done this, which moves, how many times have you done this?

Betsy Hoffman Well, [laughs] first of all, moving from Northwestern to Purdue. And then when *Matthew* [*Spitzer*] and I started at Northwestern, we ran half our experiments in the economics department and half our experiments in the law school. He would run half of the experiments in the economics, and I would run half the experiments. We would do the same thing in the law school, so that if there was any difference in subject pools or any difference in experimenter, it would come out in the repetitions.

Reinhard and I were talking last night, and he said, "You always wanted more repetitions, more repetitions," which is true. I always wanted more repetitions so we could get better statistical analysis, and also make sure that you could replicate [results], if you had multiple experimenters at multiple locations. Then I moved to Purdue, and Matthew moved to USC, and we did the same thing. We replicated the experiments half at Purdue and half at USC, and then we would trade places. And I would run half of them at USC. We would have four sets of experiments to each run with the other person at the other place.

Chris Starmer And can you recall any particular difficulties in replicating the results that you were trying to presumably get as your baseline in some sense?

Betsy Hoffman Well, in the original Coase [theorem] experiments, for example, the law students spent more time talking. They didn't necessarily come to a different outcome, but they spent a lot more time talking.

John Ledyard Billable hours. [some laughter]

Betsy Hoffman Billable hours, right. I trusted the results more, I would say, by doing that kind of careful replication.

| | |
|---|---|
| Chris Starmer | Vernon. |
| Vernon Smith | I just want [to make] a short comment on when we found that earning the right to a privileged position made a difference. We didn't go far enough. [*Jason*] *Shogren* came along and said you got to earn the money. Now it is no longer the experimenter's money. There is a sense in which the money was the individual's. And that does away with all the dictator problem. Now, people essentially give zero in the dictator game when it is their own money rather than the experimenter's money, and I think that raises questions about payoffs that hadn't been raised before. [Hoffman: yeah]<br>And it raises the experimenter's money problem rather than just whether the rewards are good. We have no idea, I think, yet what the full implications of that might be. Almost certainly, for many experiments, it will make no difference at all. But if you have got one counter example where it makes all the difference, you have got to ask: "Whoops, how far does that go?" |
| Charlie Plott | Is this published, Vernon? |
| Vernon Smith | What is that? |
| Charlie Plott | I haven't seen this result. |
| Frans v. Winden | No, I haven't seen it. |
| Vernon Smith | Oh yeah. That is the end of the dictator game. |
| Charlie Holt | What is the title of that paper?[24] |
| Vernon Smith | It is in the AER. |
| Charlie Holt | It has a really catchy title. |
| Vernon Smith | Yeah, it is a |
| Betsy Hoffman | Is this Shogren's paper? |
| Charlie Plott | If it is your paper in the AER, I will find it. |
| Vernon Smith | But that is being explored now. |
| Betsy Hoffman | No, it isn't our paper. It is Shogren's paper. We didn't do that in our paper. |
| Vernon Smith | No. We found [earned rights] made a difference. He is [Shogren &Cherry] finding that if it is basically your own money, it makes all the difference.[LXIX] |
| John Ledyard | Is this the Cherry Shogren? |
| Charlie Holt | *Hardnose the Dictator*. |
| Vernon Smith | Yes, Cherry Shogren and some one else. |

[24] **Cherry, Todd L.; Peter Frykblom and Jason F. Shogren.** 2002. "Hardnose the Dictator." *The American Economic Review*, 92(4), 1218–21.

| | |
|---|---|
| Betsy Hoffman | But then Ball and *Eckel* or *Eckel* and Grossman found that if it was given to a charity, they [the proposers] would give it all away.[25] LXX |
| Vernon Smith | Yes. Things like that, and this is back to your [Betsy's] point though. You start with the assumption that anything can make a difference. |
| Betsy Hoffman | That is right. |
| Vernon Smith | And it is a really good starting point because you will be less surprised. |
| Chris Starmer | John [Kagel]. |
| John Kagel | The dictator game is just notoriously unstable. I mean, that is more what we found out over time. It is just notoriously unstable. |
| Betsy Hoffman | But I think we have learned a lot from the things we have done to alter it. From the unstable outcomes you can make them predictably unstable by what you do, by anonymity, by making people use their own money. |
| John Kagel | I'm not sure I agree with that, but I will let it go with that. But if you review that work, it is just incredibly unstable. |
| Chris Starmer | Charlie [Holt], last comment. |
| Charlie Holt | There was a question about—your question was about how much detail is necessary. My approach is I do all of my own programming. I think when you see the details of the code, the way the market is clearing, you learn a lot. And then when I'm in the lab and I read the instructions, I will spot things that aren't right. Also, being in the lab is really important. [For instance, once] we found that some subjects who participated in an experiment come back. They might get their roommate to sign up, and then just come show up and give you the wrong name. So we started checking ID cards. Someone on the spot would spot that. But if you just pay a lab assistant and let them run, you might not catch things like that. [Smith: ehm] [Hoffman: absolutely] |
| Chris Starmer | When did you start checking ID's? |
| Charlie Holt | We ask them to show an ID when they sign in, so we have the list of people, they show you the ID. |
| Chris Starmer | How long ago did you learn that you needed to do that? |
| Charlie Holt | That was this spring. We were running some very high payoff experiments where they would earn $600.00 or something, so maybe that was bringing them out. [Ledyard: yeah; group laughter] |

[25] **Eckel, Catherine C. and Philip J. Grossman.** 1996. "Altruism in Anonymous Dictator Games." *Games and Economic Behavior*, 16(2), 181–91.

| | |
|---|---|
| Al Roth | John [Kagel]'s comment about the instability of the dictator game reminded me of John Ledyard's chapter in the 1995 *Handbook* where he talked about his experiences as a physics undergraduate. That if you painted a Ping-Pong ball silver, it looked like a steel ball, but you could make it behave differently by blowing on it. I think some of the questions about the dictator game is very much like that ping pong ball. There is much less instability say in an ultimatum game where just in the structure of the game, there is another player who acts to limit the freedom of the first mover. I think that one of the things we learned about something like the dictator game is that it is like a painted Ping-Pong ball. |
| Betsy Hoffman | On the other hand, lots of people have replicated the double anonymous results in the dictator game. |
| Al Roth | Well, no, no. But the Ping-Pong ball is also very replicable. |
| John Ledyard | Yeah, that is replicable. |
| Betsy Hoffman | Yeah, but in terms of instability, you can replicate the instability. |
| Charlie Plott | I don't understand that now. |
| John Kagel | No, what I meant by it being unstable is that very small changes can have a very large impact. That is all. If you repeat the same [treatment]– |
| John Ledyard | Delicate is the word I like to use with these things. |
| Betsy Hoffman | But it is replicable. |
| John Kagel | Yes. By whatever treatment you are using. Exactly. |
| Chris Starmer | Well, on that note, to small changes having a big impact, I propose we move to one small change that we take some coffee and we see what impact it has in about half an hour. [some laughter] Thanks very much. |

## How to Ask an Experimental Question?

| | |
|---|---|
| Chris Starmer | Welcome back. I wanted to continue exploring the theme of knowledge and skills. Towards the end of the last session, Charlie [Plott] started talking about ways in which he aims to train people to learn about experiments: teach them what an experiment is, how to ask a question. I wonder whether we could explore that some more. I wanted to know different people's experience with how knowledge about experimentation is transmitted within institutions that you |

| | |
|---|---|
| | have been in and how that has changed over the years. And I wonder if anyone would like to start off that. Al. |
| Al Roth | One thing that has changed is there is now a lot of experimental literature. Now when you teach students about experiments, there are lots of papers they can read. When you ask them to suggest experiments, they will often suggest experiments based on experiments they have read about. That is a very different kind of conversation than experiments on something novel that they are interested in. I think that as the literature gets richer, the kinds of experiments that people do, are sometimes becoming more academic. |
| Chris Starmer | How did it work before there was so much literature? Going back some time, when there was a much more narrow literature to draw on, what were the strategies then? Betsy, then Frans. |
| Betsy Hoffman | I can talk about my experience working with Charlie [Plott] and it is exactly what he was describing. He had a seminar, I don't know if it is still taught spring quarter, but it was taught spring quarter.[26] The assignment in the seminar was you had to design two experiments and run pilots. He had a little bit of money that he would allow you to use to run a pilot experiment. In the early days, it was literally learning by doing. [Smith: ehm] It was, as he just described, you proposed to him: "I wanna run an experiment on such-and-such a topic," and he would quiz you and quiz you until you got to something that he thought was actually doable, that was sufficiently defined that could actually be an experiment.<br>Then, you had to write up the instructions, and then he would critique the instructions. And then you had to run your first pilot experiment. And then he would critique the way the first pilot experiment was run, and you would run a couple more pilots. Then, you would go on to the next paper. At the end of the semester, the output of the seminar was the data that you had generated from the pilot experiments. And in my case, both of those turned into published papers with Charlie over time. I think in the early days, that was really the strategy, but it sounds like that is what you still do. |
| Charlie Plott | More or less. Now, much more of it is high tech stuff. |
| Betsy Hoffman | Sure. |
| Charlie Plott | Now, I still try to get all of the students to do something by hand first, take something that is out there like a double |

[26] Plott has been teaching a course on Experimental Methods of Political Economy since Fall 1973. Currently it is a three term (one academic year) course and is typically taken by second year graduate students at Caltech, but also undergraduates.

auction or any committee—anything that is quite simple that can be done by hand because I think that—I still think that doing them by hand is really an important thing. You have to get the feeling for what is happening out there. Then, they will typically—now, we try to move them on to more high tech stuff. We have programs, for example, we have developed a system [that runs on] Ubuntu.[27]

Take any of my complex experiments. You could take a thumbnail, put it in a computer anywhere, and it will load my machine. Download that thumbnail and it will create that virtual machine that you could use as a server. It will take all the programs off of my machine. So, you really have my machine in a virtual machine on your thumbnail. You could run your experiment, it will send the data to a Gmail account, and put it on your thumbnail. But now, that is a pretty high tech thing. [Starmer: Yes]

You have to learn how to do that. That, itself, takes down quite a bit of interaction with these guys. So, they will learn that. Once they have that, though, then, their lab is anywhere that they are with a machine that is connected to the internet.

Chris Starmer: Can I put on hold a little bit the lab stuff.

Charlie Plott: Yeah.

Chris Starmer: I see the connection there from teaching to lab, but I don't want to move too much into the lab just yet. Frans, you wanted to come.

## Designing an Experiment Is a Joint Effort

Frans v. Winden: The discussion part that Charlie was talking about was certainly important in the beginning, when we started as CREED at the beginning of the '90s, and still is. But something that comes before and something after is also important. Now, there are courses in experimental economics at the undergraduate level and at the graduate level. So, typically people that come as Ph.D. students have been through such a class. They know already how to design experiments, what an experiment is—they have participated in some experiments. Then you get into this intimate discussion part where you try to figure out a good question.

Chris Starmer: People's experience. Now that people probably tend to come as graduate students thinking about running experiments, but

[27] A Linux-based operating system.

with much more prior knowledge of literature. Do they come with better questions?

Frans v. Winden They are much more informed. They know that it is important to come up with something fundamental, not a complex issue—like Charlie [Plott] said. They have heard about it now in these courses. That helps a lot. But what I wanted to add is related to [what comes after]. You have this discussion part, and then they start working on designing an experiment and how to deal with software packages. This is followed by two other important events. That is, first of all, the CREED lab seminar where they present the design. We work as a research group. We have four professors, so there are quite a few students.[28]

[In principle,] all are attending these meetings because they all rely on good feedback about the design. Everyone feels responsibility and commitment to attend these meetings. So, students get a lot of feedback. Also external people who would like to use our lab are required to participate in these lab seminars and present their design. Then later we have the pilot seminar where software and instructions are tested. Regularly, people have to rework their design because something shows up that is still worrisome. Now typically people go through courses, then have discussions, lab seminars, pilot seminars, and then start doing experiments.

Chris Starmer What sorts of different models operate at different institutions?

## Learning from Teaching

John Kagel We don't have that degree of continuity because our graduate students are typically not our undergraduate students. There is a big difference. Occasionally, if I do an experiment with an undergraduate, a thing that I have found to work quite well is to take one of my more experienced graduate students and have them work with the undergraduate. They form a team and they come to you when there are any particular issues. We teach or at least historically we have taught, at Ohio State,

[28] In 2010 the full professors at CREED were Frans van Winden, Arthur Schram, Joep Sonnemans, and Theo Offerman.

experiments in the context of a second year course in the Industrial Organization sequence where we cover auctions.

*Dan* [*Levin*] and I teach a course covering auctions. Half the course is spent on theory, half is spent on experimental and a little bit of empirical work in that. We do ask them to establish a research proposal on the basis of the course. And we have been quite flexible in the sense that we don't require that they do it any particular—in auctions because we know that they are not all going to be interested in that particular topic. Our better students have often used that as the basis for their third year paper.

Chris Starmer: When did you start doing this model?

John Kagel: Right when I got to Ohio State, I guess, in 1999, *Dan [Levin]* and I started to teach the course. And then, I think, the year after that, and we have been using that model ever since. We have been pushing for in this class, instead of having—or within the Industrial Organization sequence, instead of having a field comp[rehensive exam] to have them write a paper and get some resistance from our colleagues on that.[LXXI] We haven't changed the rules yet, but we are pushing the rules. [Hoffman: That's good]

Chris Starmer: I'm interested that you deliver training in experimental economics via a module on IO. Why do you do it that way rather than, for instance, having explicit [unfinished].

John Kagel: Well, the original idea was that experimental work is a method applied in a subject matter.[29] That was the original idea. And I think it still goes back to what we were talking about yesterday that how—I tell my students, if you go on the [job] market and they will inevitably ask you what can you teach? And if your answer is experimental economics and that is it, I say, "You are dead. You can't get a job because there are just not enough courses like that." We get paid for teaching subject matter courses and they have to be able to say they can teach something at the graduate level that is above just experimental economics.

Chris Starmer: But it is not clear that having done a module in experimental economics would be a hindrance to

John Kagel: No.

[29] **Editor:** Could you please elaborate a bit more on this "original idea"? **John Kagel**: "I meant to say experiments are a methodology applied to problems in different subject areas. That is, it's not like econometrics where there are significant mathematical problems to be solved that are of general use—theorems to prove. With experiments one solves methodological problems as they arise in a context/application. These may wind up having general application—but it is a far cry from solving math theorems about convergence to asymptotic states and unbiased estimators."

Chris Starmer: Doesn't seem to explain that decision to deliver the course in that way.

John Kagel: The explanation for delivering it in that way is that it was a natural way within the existing sequence. I should add that now that we have hired an experimenter from Al's group,[30] we are going to teach a stand-alone course in experimental methods, which will be more like what Charlie [Plott] is talking about. It will be a course they can take in a third or fourth year.[LXXII]

Al Roth: So, at Harvard we have a couple courses in experimental economics, we have an undergraduate course that is called Economics and Psychology that *David Laibson* teaches. We have . . .

Chris Starmer: When was that introduced?

Al Roth: He has probably been doing it for ten years with a graduate course in experimental economics that I teach that is about the things we talk about here. There is a course in field experiment . . .

Chris Starmer: Again, if I may, when did you introduce that?

Al Roth: I started that when I came in '98. There is a course in field experiments that someone like Sendhil Mullainathan and Nava Ashraf will teach about doing randomized trials in the field.

Chris Starmer: Is that newer?

Al Roth: It is newer. And there are—there is probably an experimental course taught by Jen Lerner in the Kennedy School. They have a new lab[31] that has lots of physiological tools in it, and I think they give a course of their own. Harvard is a little bit decentralized. [Betsy: a little bit? Some laughter] Different parts of the university have different courses, but they are all—and sometimes students take more than one of them. I see my students who have taken one of the other courses.

Chris Starmer: Charlie [Plott].

Charlie Plott: I was going to follow up with that. I was talking about how I teach my course. My course is typically about markets and market-related stuff. There are other courses at Caltech, for example, in game theory. That is a different experimental literature. Then, there are still other courses that are taught on individual choices that typically go with biology, FMRIs, and neuroscience. They also take a lot of biology and a lot of neuroscience. So, they are taught with neuroscience graduate students and post-docs.

---

[30] That was Luke Coffman.

[31] It is the Harvard Decision Science Laboratory which opened on December 5, 2008.

| | |
|---|---|
| Chris Starmer | Can you say something about the timeline in which these different courses . . . |
| Charlie Plott | When were these courses introduced? |
| Chris Starmer | Yeah. |
| Charlie Plott | Oh well, of course, my course has been around since the beginning of time. [some laughter] |
| John Ledyard | Before. |
| Betsy Hoffman | '70s. '73 or '74. |
| John Ledyard | Before |
| Charlie Plott | Informally '71 perhaps. The first time it was really taught was when Vernon and I taught it in, what, '73 maybe? |
| Vernon Smith | Yeah, it was in the Spring [quarter] of '73. Yeah, that was one of those milestones that we mentioned the other day.[32] |
| Charlie Plott | Yes, that is a milestone. |
| Vernon Smith | Kind of a watershed point—certainly for me and for Charlie [Plott], and it got experiments started at Caltech. We jointly taught a seminar. I had taught a seminar at Purdue from 1963 to when I left in '67. So, I had notes and also reading ideas. Then, we supplemented that with stuff that had come out since. In the class, we had, I think, two or three paying customers, and the rest were the faculty. |
| Charlie Plott | Ehm. *Kim Border* and *Mo Fiorina*. |
| Vernon Smith | *Mo Fiorina*, *John Ferejohn*, *Roger Noll*, *Bill Riker* was visiting that year also as a Fairchild Scholar, he sat in that course. *Jim Quirk*. There were seven of them. I have left someone out.[33] And that was basically it. These guys all introduced experiments in whatever they were teaching at Caltech, right? |
| Charlie Plott | Uh-huh. |
| Vernon Smith | Huge synergies came out of that. I remember *Bill Riker*. He had been commissioned to write a paper on experiments in political science.[34] Do you remember this? |
| Charlie Plott | Yeah, of course. |
| Vernon Smith | And after the end of that quarter he wrote the guy back. He said, "I need more [time to finish]. I have learnt so much. It has changed the way I thought about it." It really affected all of our thinking. |
| Chris Starmer | Can you say something about how? |

[32] Smith was a Fairchild scholar at Caltech in the 1973-74 academic year. So the course took place in the spring period of 1974, not 1973. He stayed in Pasadena in 1974-75 as a visiting professor at Caltech and USC, and taught a course at Caltech that academic year.

[33] The identity of the seventh attendee remains unknown. Smith like Riker visited Caltech through the Fairchild Scholar Program.

[34] This survey paper has actually never been written.

| | |
|---|---|
| Vernon Smith | How? Well, Charlie will have to speak for himself. He had had these experiments with [*Harvey*] *Reed* and he had also started to do the public goods experiment. I didn't know about that, but he had already started. |
| Charlie Plott | I had already done committee and agenda experiments. |
| Vernon Smith | But we did not cover that in the seminar. [Plott: Yeah] And because the seminar mostly dealt with the early stuff—the work that had been done and variation on market rules, wouldn't you say, Charlie? |
| Charlie Plott | Uh-huh. |
| Vernon Smith | That was the primary emphasis. We didn't get into any of the new topics that Charlie was—apparently had been working on and was working on at the time. I think the synergy came from people who—*Bill Riker* is another person who had had some experimental [experience]—he's been doing political science stuff for actually years, hadn't he?[35] |
| Charlie Plott | Yes. And *Mo Fiorina* was his research assistant. |
| Vernon Smith | Yeah, and *Mo Fiorina*. And then, there were the others that really were learning this for the first time, you see, and were highly motivated, were really interested in this topic. That, as I look back, was an important part of the synergy. Then those—like *Roger Noll* got into doing experiments after that and so did *John Ferejohn*, but he hadn't done any before. Is that correct? |
| Charlie Plott | No, neither of them did. There were some really interesting discoveries. The concept of efficiency came out of that seminar. The posted price effect came out of that seminar.[36] There is just this whole series [Smith: Yeah] of really important things that lasted since then. The role of posted prices and again the institution structure of how that affects market convergence came out of that seminar. Corrected versions of asymmetric information came out of that—whole series of really interesting discoveries that were so basic at the time, now people have forgotten about them because they are so old. |

[35] **Riker, William H.** 1967. "Bargaining in a Three-Person Game." *The American Political Science Review*, 61(3), 642–56, ____. 1971. "An Experimental Examination of Formal and Informal Rules of a Three-Person Game," B. Lieberman, *Social Choice*. New York: Gordon and Breach, , **Riker, William H. and Richard G. Niemi.** 1964. "Anonymity and Rationality in the Essential Three-Person Game." *Human Relations*, 17(2), 131–41, **Riker, William H. and William James Zavoina.** 1970. "Rational Behavior in Politics: Evidence from a Three Person Game." *The American Political Science Review*, 64(1), 48–60.

[36] What Plott refers to as the concept of efficiency and the posted price market institution were introduced in a joint paper with Smith: **Plott, Charles R. and Vernon L. Smith.** 1978. "An Experimental Examination of Two Exchange Institutions." *The Review of Economic Studies*, 45 (1), 133–53.

But at the time, they were quite important. Especially that concept of efficiency, which, of course, everybody uses now and never thinks about where it came from. Well, it hadn't been around all the time. It came out of that seminar.

**Vernon Smith** Yeah. Learning by doing as a teaching device was the way I taught. I didn't know any other way to teach experimental economics from 1963 to '67 other than involve the students not only in the literature, but coming up with the project, and then doing it. It seemed to me—what else was there?—we had no tradition. We were making it. We were learning what the tradition was. And it made for pretty exciting teaching because it meant the students were learning and you were learning as much as the students. [Vernon laughs] And a lot of those experiments didn't get published, that came out of the—that the students were doing from '63 to '67, but it was a huge influence—we talked about them at every [class] meeting what the projects were. Other students were learning from the experiences of any one student who was discussing his work on any given day.

**Chris Starmer** Charlie [Holt], it seems to me you have been very closely involved in developing approaches to teaching experiments. I wonder if you could say a bit about your approaches and perhaps how those have changed over the years.

**Charlie Holt** Well, when I teach undergraduates, I divide them into teams and they design their own experiment.

**Chris Starmer** How long has this been part of the undergraduate program?

**Charlie Holt** Since about '88.[LXXIII] I don't know. But anyway, what they do is they—they are not used to working in teams. They get together and design an experiment, and the web-based programs that I have give them lots of choices.[LXXIV] But many times, they do them by hand as well. Then, they will go home and do a PowerPoint and present that at the beginning of the next class. The students in the class who had been in the experiment, they do the reading the same time the team is preparing the PowerPoint. They have to do the reading because they do an online quiz.

So, they walk into class the next time and they have been in an experiment, they have done the reading, and they are seeing their classmates do the presentation. I learn a lot from those myself, and sometimes the research experiments are tried out in the classroom. For example, I got *Lisa Anderson* to come watch the first information cascade experiment I did in undergraduate class,[37] because I told her it was really interesting.

Chris Starmer: And when was this?

Charlie Holt: This would have been maybe '93 or '94. Anyway, it was interesting because there, you get a signal, and you make a prediction as to which urn the marbles are being drawn from. You could tell a student, if someone said, "Oh, I think it's the blue urn," and someone else said, "I think it's the blue urn." Someone else gets a signal, and they hesitate. You can tell they must have gotten the contrary draw. And so, the fact that everybody was looking at them made me realize that when we do these as research experiments, we have got to be very careful and isolate them really well.

From then on in the class, we take them out of them room, and draw it from out of the room, let them predict out of the room, and come back in and announce it. [The class experiment] helped me to be more careful with the research experiment. It helped me realize that there are some things you are going to have to be more careful about when you run [an experiment.]

Coming back to the idea of being involved. I know we did some things on the game that *David Reilly* has written about and he called stripped down poker.[38, LXXV] You get a draw and you decide whether to raise, and the other person can call your bluff.

And the same thing happened with computer stuff, people would realize if there was a time delay, then, it would have been that they drew a weak card, and they are just thinking about whether to bluff or not. When we did those as research experiments, we had to rig it so that you couldn't tell anything from the delay. Everybody had to wait until all decisions had been made before they got the results.

Chris Starmer: When did you discover that you needed to do that?

Charlie Holt: That would have been about eight years ago. But it was from the computer code. If you let them go at their own pace, they got a quick response. "They figure all that—they raise real quickly. They must have an Ace." And so, they were inferring something from the timing that if you just wrote the computer

---

[37] **Anderson, Lisa R. and Charles A. Holt.** 1996. "Information Cascades." *The Journal of Economic Perspectives (Nashville)*, 10(04), 187–93, ____. 1997. "Information Cascades in the Laboratory." *The American Economic Review*, 87(5), 847–62.

Holt had an NSF grant (SBR-9320617) *Information Cascade Experiments* during 1994–96.

[38] **Reiley, David H.; Michael B. Urbancic and Mark Walker.** 2008. "Stripped-Down Poker: A Classroom Game with Signaling and Bluffing." *The Journal of Economic Education*, 39(4), 323–41.

code and ran the experiment without debriefing the subjects, without talking to them about what they were doing afterwards, you might get some funny results in the research experiment. So, debriefing's important, too.

## Thinking as an Experimenter

Chris Starmer: Before these students get to these projects, I'm guessing they have been to some classes where you have introduced them to experimental methods. What things do you want them to know before they start on these projects?

Charlie Holt: Well, I guess you need to do a few decision experiments, a few games, a few markets. But if you teach them too much, then, they are—what they think of is too closely tied to the literature.

Chris Starmer: It sounds like it is okay—they are getting their hands dirty by learning something about the literature rather than starting them off with learning about experimental design, in the abstract

Charlie Holt: That is true.

Chris Starmer: like randomization. It is to introduce them to literature. Is that . . .

John Kagel: At Pittsburgh[39] we [referring to Al Roth] had an experimental class in the undergraduate program. The reason you don't start off with, at least I don't start off, with abstract things is that the key issue in terms of designing an experiment is what is the core behavioral issue that people could mess up on or have to grasp in order for the theoretical prediction to come about? And you see that almost by example of other established experiments where people have narrowed that down, and [identified] what is the core issue.

But that is what starts off as the experimental idea. Then, the question of how should you have good procedures and things like that all follow once they have the fundamental idea. That has been my approach.

Al Roth: There is now so much experimental literature that gives lots of students an entry. Typically, my experimental economics course has a mix of first year and second year graduate students. What I have found is that when they start to propose what experiments they will do, first year graduate students typically propose they will do some experiment related to an experiment that we have talked about in class.

[39] Kagel taught at Pittsburg from 1988 until 1999.

Second year students often propose they will do something, an experiment related to something they were already interested in when they came into the class.

I think my first course was probably when I was at the University of Illinois.[40] One thing that is different today than when we started teaching experiments is that it is a great benefit for experimentation as a field that we now have all this depth of history of experiments you can do things related to other people. On the other hand, I think I would regret if it turned out that experimenters were people who studied particular things. If experimenters were people who studied ultimatum games and double auctions as opposed to if they were economists who use experiments to study whatever they are interested in. So, at some point, I thought of limiting my class to only second year graduate students.

I talked about it with the class, and the first year students said to me, "Maybe it is more fun for you to read the proposal of second year students, but those of us who take it as first years, we will go into our second year courses knowing that experiments can be done, and we will take our field courses with a different point of view than if we hadn't had the experimental class." So, I have left it open for people to take it when they like.

Chris Starmer — You mentioned the first training that you got in experimentation. I think you said in Illinois?

Al Roth — Well, so, first training was when I met *Keith Murnighan*, and he and I taught each other how to do experimental economics. He already knew how to do experimental psychology.

Chris Starmer — I am interested to hear more about people's own first engagements, when they were at the learning end of this. For instance, Betsy, you said yesterday that, I think you were in Charlie [Plott]'s class, this was a life-changing experience.

Betsy Hoffman — Yes.

Chris Starmer — Can you say more about that?

Betsy Hoffman — You want me to elaborate on that? I had been headed down a path of being a historian for a very long time. I majored in history at Smith [College]. I wrote an undergraduate thesis on the Black Death, and then I went on to graduate school at the University of Pennsylvania, and I continued to study medieval and Renaissance history, but also demography. I got a minor in demography, and ended up writing a thesis on the declining mortality in Italy, which actually, people still read, which is amazing.[41] But I got into a dead end in a sense that the job

[40] Roth worked at University of Illinois from 1974 until 1982.

[41] Hoffman defended her history dissertation in 1972. For details see Chapter 2, Footnote 47.

market was horrible in history and I ended up at the University of Florida, where I was the first woman ever hired.

I was hired under threat of a suit under Title Seven. I had just gotten a divorce, and nobody understood a word I had to say because I think in terms of formal models. Actually, *Andrej* [*Svorenčík*] and I, at dinner last night, talked about the difference in the way we think. And he thinks like a historian, which is what got him interested in this subject [history of experimental economics]. I think like an economist, but I didn't know I thought like an economist. Because [when I came to Caltech] I had only had principles of economics as an undergraduate and worked only with *Richard Easterlin*, who is an economist. But I had worked on a demography thesis. So, I really didn't think of myself as an economist.

I got the opportunity to go to Caltech really as a post-doc through *Lance Davis*. But since they didn't have a post-doc program, he gave me a fellowship to come to Caltech as a graduate student, but neither of us ever expected I would get a degree. Both of us viewed this as a post-doc and my view was that I was going to gain the mathematical, the modeling, and the statistical skills to go with my basic thinking, and my basic ability to deal with data, and my basic ability to think in terms of formal models. I was going to put a structure; a mathematical, theoretical, and econometric structure on the way I had been working anyway.

And I always assumed I would go back into the history department. Then, I was convinced that I would be better off, I would be more accepted as an economic historian than in an economics department. At that point, I decided to finish my Ph. D. in economics, but my intention was always to be an economic historian until I took Charlie's course. As I said yesterday, it was so much fun. It wasn't that I hadn't had immense amounts of fun being an economic historian, and I still for probably another ten years continued to do work in economic history. I have worked with *Joel Mokyr*, I have worked with Ann Carlos. I became the modeler. After being at Caltech, I got recruited by economic historians to run models. But it was working with Charlie that—it was just so much fun.

Chris Starmer I don't doubt that it was fun, but I'm sure it was more than fun.

Betsy Hoffman It was more than fun because I have always been a maverick. I have always done—I have always, my entire life, from the time I was a child, done things that were different than anybody else. And here was this brand new field, coming up with brand new insights, using brand new techniques, and it

| | |
|---|---|
| | was that that excited me. [Starmer: Right] It was being—just as I was at the forefront of the transition of history into mathematical, statistical, and quantitative thinking. What was interesting to me was to be at the forefront of a new field. And just before I really gave up full-time research, I was moving into neural economics.<br>To me, I'm always interested in what is new, and exciting, and what is driving the discipline. I get bored when it is just regurgitating old stuff. I want to do something that breaks the mold. That is what is fun to me. |
| Chris Starmer | Yeah. John [Kagel]? |

## Transferable Skills from Other Fields

| | |
|---|---|
| John Kagel | Well, you posed a question of how did some of us learn to do experiments when there were no experiments in economics to look at. In my case, it was from working with *Rachlin* and *Green*, the operant psychologists. They had this whole bunch of techniques. |
| Chris Starmer | When was this? |
| John Kagel | Oh, that would be when I did the first animal experiments. That would be in the early, mid '70s. And the same thing when working with the operant [psychologists] who work with people in the token economy. They have a whole literature on experimenter demand effects and designs using—going between groups versus within group designs and the issues that go along with that. We just were taking their experimental methods and applying them to economics. |
| Chris Starmer | It seems like that to some extent token economies and in particular animal experiments draw on a set of skills that are atypical of experimental economics. That must have ... |
| John Kagel | Those are the detailed lab skills. They used these switch wires. If you were a good electrician, you could program one of these things. Turns out that [*Ray*] *Battalio* had been an electrician in the Navy. So, he could program these things when we set up our [animal] lab. That was just lucky. Then, *Len Green* came down and helps to set things up and shows Ray how this stuff works. After, Ray could deal with that. Winds up that he had a neighbor who was a vet. So if you have a rat that is doing something really weird, and you want to know is it something you did to them or is it because there is some disease, we had the neighbor come in and tell us what was going on with the rat. We were just lucky with those particular skills, and that Ray had been an electrician in the Navy. |

Vernon Smith: But John, you did a transition from that to humans.

John Kagel: Yes.

Vernon Smith: And that—those skills don't necessarily carry over. [some laughter]

John Kagel: I didn't have those skills. In terms of the design skills—how to design an experiment—those skills carry.

Vernon Smith: Yes.

Chris Starmer: And was the transition straightforward or . . .

John Kagel: Oh yeah, the transition was straightforward. In fact, in terms of going to University of Houston, I had to promise them I would do something with humans. [laughter] This was my commitment. I brought people in. Vernon came in. He talked about his private value auctions.

Chris Starmer: When was this?

John Kagel: Boy, that must have been the first or second year I was at the University of Houston. That would be '82, '83.[42] And he [Vernon] comes in. And there is *Wilson's* beautiful paper about how the common value auction, as the number of number of bidders goes to infinity, you are going to get the right answer. [i.e. the price will converge to the true value of the item][43] [Smith: ehm] There was this younger guy from MIT who's an assistant professor there. I say, "Are you interested in looking at this, and doing an experiment on this? We could do something on common value auctions. No one has touched that area." And that was [Dan] *Levin*. He said, "No, no, I don't do experiments." So, I said, "Read Wilson's paper."

Then he read Wilson's paper, and he tells me to this day: "You just sucked me totally in." He was just so taken by Wilson's paper, and we designed the experiment from that. But we had Vernon's particular techniques used in private value auctions that could be adapted to common value auctions. I knew the design things already, and could apply certain things that were a little bit unusual at the time.

Al Roth: There is a different class of skills that we have been talking about that John has just touched on, which has to do with what are the experimental treatments you want, how that depends on what you are testing, and what audience you are speaking to. The thing about *Dan Levin* is John worked with *Ray* [*Battalio*], and Ray was an electrician and programmed machines. Dan was an auction theorist and could start to talk

---

[42] Kagel moved to Houston in 1982.

[43] **Wilson, Robert.** 1977. "A Bidding Model of Perfect Competition." *The Review of Economic Studies*, 44(3), 511–18.

to the community of auction theorists. And the question is how should you design experiments, what should you control for.

John Kagel — He became an auction theorist.

Al Roth — Yeah. Yeah.

I originally had a very false model of how experiments might proceed. One of my earlier experiments on bargaining, we published it in *Psychological Review*, which is a journal like the AER for psychologists.[44] And my model had been—there had been a little literature in the psychological journals about trying to test economic models of bargaining. And by and large, economists were really unimpressed with that because it didn't control for the things that they thought should be controlled.

And so, I thought: "We will publish a paper in a psychology journal showing how to control for the kinds of things that my game theory colleagues were worried about, and then, psychologists will carry the ball forward." [laughter] But of course, that didn't happen because psychologists aren't interested in testing the kind of things that economists are interested in. So, we had to carry it forward ourselves. And we moved into economics journals so that people wouldn't think, "Why are you testing these odd theories?"

I actually tested a set of bargaining theories that performed very badly and have, more or less, stopped being the standard to use in bargaining. But I had got a lot of credibility among theorists because I had done a lot of that theory. Sort of like Charlie [Plott] in social choice, I had written a book called *Axiomatic Models of Bargaining*.[45] And then started to bring those things into a laboratory. But how to bring them into a laboratory so that you could show the theory community what was going on, that is a set of skills in how to—in which experiments to do that are different from the skills of how to recruit people into the laboratory, make sure they don't talk to each other, and all of that.

I think that different kinds of experiments require different skills depending on what it is you are looking for and who you are talking to about it.

Chris Starmer — You focused that around theory testing and what were you thinking of as different kinds of experiments? Are you thinking of theory testing being one kind, some other experiments being . . .

[44] **Roth, Alvin E. and Michael W. Malouf.** 1979. "Game-Theoretic Models and the Role of Information in Bargaining." *Psychological Review*, 86(6), 574–94.

[45] **Roth, Alvin E.** 1979. *Axiomatic Models of Bargaining*. Berlin; New York: Springer-Verlag. Available online http://web.stanford.edu/~alroth/Axiomatic_Models_of_Bargaining.pdf [Accessed on March 31, 2015].

## The Questions Depend on the Relevant Audience

Al Roth No, no. So, theory testing depends on theory. Being familiar with the theory is helpful and different theories will require different kinds of treatments to elucidate what is going on. But also in dealing with practical matters. For instance, I have designed clearinghouses of various sorts. At some point, John [Kagel] and I did an experiment that aimed at elucidating the success and failure of various clearinghouses in markets for British first employment doctors for house officers. That was part of a discussion I was having with American medical administrators.[46]

The thing about that experiment—people had done experiments before focused on clearinghouses and [investigated] whether stability was important, but we were looking at the experiments in a context where they seemed to make a lot of sense. That is, in the context of first allowing a decentralized market to experience a certain kind of market failure, unravel this market failure, and then trying to fix it with a centralized clearinghouse. The fact that we did that experiment in that context made it germane to the discussion we were having about market design with that other experiments that simply looked at static environments hadn't been germane to. So, understanding something about those markets helped us design experiments that addressed issues in those markets. [47]

Chris Starmer Charlie [Plott].

## Learning the Theory from Experiments

Charlie Plott I was going to pick up on his comment about designing the experiments and I just pass along an observation. Almost every time I broke into a new area like asymmetric

[46] **Kagel, John H. and Alvin E. Roth.** 2000. "The Dynamics of Reorganization in Matching Markets: A Laboratory Experiment Motivated by a Natural Experiment." *Quarterly Journal of Economics*, 115(1), 201–35.

[47] Roth is referring to: **Harrison, Glenn W. and Kevin A. Mccabe.** 1996. "Stability and Preference Distortion in Resource Matching: An Experimental Study of the Marriage Market," M. R. Isaac, *Research in Experimental Economics. Vol. 8.* Greenwich, CT: JAI Press, **Sondak, Harris and Max H. Bazerman.** 1989. "Matching and Negotiation Processes in Quasi-Markets." *Organizational Behavior and Human Decision Processes*, 44(2), 261–80, **____.** 1991. "Power Balance and the Rationality of Outcomes in Matching Markets." *Organizational Behavior and Human Decision Processes*, 50(1), 1–23.

information in the mid '70s that was with *Louis Wilde*. He was interested in search theory and asymmetric information. This had to do with professional diagnosis and there was also a funding contract there. That evolved into rational expectations when I was at the University of Chicago.[LXXVI] I teamed up with *Shyam Sunder* who was an accountant, but also in finance, who was deeply interested in rational expectations and information aggregation. Those experiments were designed for testing theories of *Lucas* and Fama, who by the way, were in the audience the first time we presented our results.

On each one of those [experiments] we expected negative results. The design of the experiment was that you didn't just test the theory because you figured the theory's just plain wrong. The question is, how bad is it, where is it going wrong, and is there an alternative that is better? So, it is not just the test of the theory, it is the alternative theories that are going to go into the design bag with it for the purpose of testing. All those things are really crucial. When we try to talk about asset markets that comes from *Bob Forsythe* who was a student of *Dave Cass* who has spent his career worrying about asset markets, or the studies of capital assent pricing markets, [that comes from] *Peter Bossaerts*, again a theoretical finance guy.

So you see that from an experimentalist point of view, breaking into a new area, [when] you are opening up a new area for experimentation, is almost always accompanied by a partner who's a really good theorist. The theory and the experimentation have gone hand in hand and it is not the case that you just test theory because that is child's play. The theories are all so naïve. It is easy to reject them. The question is trying to get the essence of the phenomenon out so that you can actually see the theory evolve and accommodate what there is of use.

Chris Starmer: Essence is an interesting word. What do you mean by "get the essence of the theory"?

Charlie Plott: Well, [unfinished][LXXVII] Let's take rational expectations. It is typically built on the hypothesis that the price carries the information, people condition on the price. You update your beliefs on the state given the price. That is not the way it happens. People pay no attention at all to the price. All the information comes from the bids and asks. If you did the [market experiment with] posted price [institutions] where the bids and asks are cut out, there will not be much information aggregation. There are these institutional details, environmental details, [Vernon: ehm] and if you just did a posted price with information aggregation, you would reject the theory, and nobody would care about the [rational expectations] theory.

[Economists] wouldn't care about any of it because it is only a simple negative result. You are not showing the theory has any life. You are not discovering any feature about the theory. Same thing, say, in Scarf models.[48] In the Scarf model, there is a unique competitive equilibrium. We also know that according to theory the prices of the system will cycle around the equilibrium. That theory is based on an abstract process like tâtonnement. But the experiments are double auctions and definitely non-tâtonnement, but cycles of the form captured by the Scarf model are observed in the data. Now we see the theory applied and successful where it has no right to apply.

[The experiment demonstrates] that the theory has teeth, it is telling us stuff. The theory's not telling us exactly what we have seen, but it is leading us down and opening up an area for which there is something to study, and the theory's catching some of it. To me saying a theory's right or wrong is not an interesting thing to do at all. The task is to say how did it go wrong, how right is it, is it better than other things? That is the essence of—and it takes a theorist to understand where those shades of judgment—the nature of those shades of adjustment. I can't do it as simply as an experimentalist.

Chris Starmer: Is it a shared view that a good experiment must have, in some sense, a well-defined theoretical background to it?

John Kagel: I think often they start that way. [v. Winden: Yeah] That they will start that way, and I agree 100 percent with Charlie [Plott] that in terms of the context of what people are interested in the subject matter of the economics or the political science that you are studying. Working with someone who has [expertise in] that subject matter is really critical. We are starting to do some legislative bargaining experiments and we have had Massimo Morelli, who is—that is his subject matter is that, and he is an integral partner in that type of work. Now, I think that once you get into these sorts of things that you start to do things for which there is maybe no theoretical...

Charlie Plott: That is true. You learn the theory.

John Kagel: Well, no, but also you might start to do a treatment that has no basis [Plott: yeah] within the theory at all. It is something that you have seen. The mechanism, the theory is working, but for a different reason. Then you might do the posted price

[48] Herbert Scarf (1930-) an American mathematical economist. Ph.D. from Princeton in 1954. He pioneered the use of numeric algorithms to solve general equilibrium systems using applied general equilibrium models.

experiment, to take Charlie's example, and find that that information doesn't aggregate and it confirms your hypothesis. But the theory never told you to do that in the first place. It was only by going into the lab and observing what was going on that it suggested you should do something like that.

Charlie Plott — Yeah.[LXXVIII]

Chris Starmer — Betsy.

Betsy Hoffman — Well, I wanted to pick up on what they were saying because a counter example to a starting with theory is the whole literature on ultimatum bargaining and dictator games. It started with a theory of what the sub-game perfect equilibrium was in this particular game, but with Reinhard Selten's early work and other people in Europe. [*Werner*] *Güth* started doing ultimatum games and getting equal split, which was contrary to the sub-game perfect equilibrium.

That spawned then a really worldwide set of research trying to understand what was going on in the ultimatum and dictator game, which then led to the development of a different theoretical structure of what was going on. And you started to realize that it is not just that the other person you are dealing with is necessarily a perfectly rational person who always maximizes individual utility and therefore will accept the smallest unit of account.[49]

Chris Starmer — But I don't see you quarreling with the idea that you started with the theory. It seems that you start with the theory, you discover surprising phenomena, and then you are on the hunt for some new theory to account for it? Does that . . .

Betsy Hoffman — [Yes,] that. But I would say by the time Vernon and I got started working on the ultimatum games, we were not interested—the theory, in some sense, in the simplest form, I don't know if disproved is the right word to say, but it had been discredited. And so, the question we were then interested in . . .

Vernon Smith — Why?

Betsy Hoffman — Why were you getting results that were—why was everybody all around the world getting results that were counter to game theoretic prediction?

Chris Starmer — But that why question you are asking, is that the same for you as saying, "What is the best theory I can come up with to explain this?" or is it something different?

Vernon Smith — Well, here it was very clear that the context was important. The theory, as it came to us, was context free. And so, the question that we were looking at was what happens if you—suppose you embed the ultimatum game and exchange

[49] The smallest monetary unit available in a particular experiment.

between the buyer and the seller. The buyer implicitly recognizes the right of the seller to go first and name price, and then, changes the—even though it is formally equivalent to the ultimatum game, it makes a difference.[LXXIX]

If he is posting a price, he is not offering as much to the other person. Is it rejected more often? No. In other words, here, people are coordinating on interpretation of that, what is mutually acceptable. You are still not getting the equilibrium of the sub-game, but the thing is, you move closer to it. So, we were just interested in exploring those …

Chris Starmer: But you are talking about, seems to me, dimensions of experiments that make a difference, you refer to context, but you seem to be talking about them as if they are somehow separate from theory, and they are separate from the theory that you start with. But are there other things one should be drawing in?

Vernon Smith: They are not. They don't have to be separate from the theory.

Chris Starmer: That is what I was unsure you were thinking about?

Vernon Smith: No, they don't have to be, but until you find out what that behavior is, [hits the desk twice] I think it is very hard to ask how it might be modeled.

Betsy Hoffman: What it led to was a deeper understanding of the interplay between individuals and the extent to which economic activity is not just a bunch of automatons playing an equilibrium game. But individuals inferring something about what the other person is thinking and making a decision based on their hypothesis about what the other person is thinking. One of the things that we discovered, for example, is when you went to $100.00, you would have thought that you would get more sub-game perfect equilibrium, when in fact, you got less.[50]

Well, in that sense, the one way to look at it is the opportunity cost of being wrong has gone up in a $100.00 game. In fact, we had a $40.00 offer rejected. And $30.00, and $20.00, and $10.00 offers regularly rejected.

## Discovering the Sub-game Perfect Equilibrium

Chris Starmer: We are getting close to the end of this session. Before we close this session, though, I …

Reinhard Selten: I want …

---

[50] That is a higher proportion of offers being at the sub game perfect equilibrium.

| | |
|---|---|
| Chris Starmer | It was Reinhard. I really wanted to hear something from you because I think you have views about these issues. |
| Reinhard Selten | I want to tell you something about the sub-game perfect equilibrium and how this came about. |
| Vernon Smith | Can you speak up? |
| Reinhard Selten | I wanted to tell you a story of how the sub-game perfect equilibrium came about. What we did first, in my early experiments, I also did experiments with situations where I didn't know or where there was no theory for it. There were many oligopoly situations, which didn't involve clear theory. For example, I looked at the situation of oligopoly with demand inertia. Demand inertia means that future demand or future sales depend on prior sales. If you have prior sales, you will have better sales later and so on in competition. And we made experiments with oligopoly inertia with demand, and investment, and all that, different interest rates for positive accounts, and for debts and so.[LXXX]<br>It was quite complex when we explored it. My associate *Otwin Becker* and I tried to make a theory for this experiment.[51] What should be the theoretical solution? Then, I think I simplified it completely. I simplified it to a high degree and kept demand in it. And I computed the equilibrium. [But] I suddenly found that this was not the only one. There were many other equilibria. And then I invented the idea of sub-game perfectness in order to single out the one equilibrium. [laughter] It was not only sub-game perfectness alone, but also asymptotic insensitivity which I required. I wrote a paper on this oligopoly model. So, actually, sub-game perfectness was motivated by experimental investigation in an oblique way.[52, LXXXI] |
| Chris Starmer | It seems you inverted the story of the [previous] example by starting with an empirical investigation and very little theoretical background and then, leading to important theoretical… |
| Reinhard Selten | Yeah and then it led me to developing a theory for a similar situation and that was it.[LXXXII] Then, later, to make an addition to the story, which was told before. We had |

[51] **Becker, Otwin and Reinhard Selten.** 1970. "Experiences with the Management Game Sinto-Market," H. Sauermann, *Beiträge zur Experimentellen Wirtschaftsforschung*. Tübingen Mohr, 136–50.

[52] **Selten, Reinhard.** 1965. "Spieltheoretische Behandlung eines Oligopolmodells mit Nachfrageträgheit. Teil 1 & 2 Eigenschaften Des Dynamischen Preisgleichgewichts" *Zeitschrift für die gesamte Staatswissenschaft*, 121, 301–24, 667–89.

psychologists in one of our conferences in Germany. I don't completely remember their names.

Chris Starmer Do you remember when or where?

Reinhard Selten Yeah, it was in one of our volumes also, bargaining.[53] This is a psychologist-led experiment where the subject played against a computer, but didn't know it was a computer. The computer was programmed to have a fixed concession rate. And they played for 20 periods. They made alternated offers and the subject was the last one to accept or reject the offer. We saw that the subject often left their $3 on the table or so. And it all resulted in a conflict. I was completely surprised about this. The psychologists were not surprised at all. [v. Winden, Hoffman laugh] They did not think anything about this, but I was surprised about it. And I discussed it with *Werner Güth* and later he made then this experiment. *Werner* simplified the whole thing to just one period to the ultimatum game, which happens at the end of this game.[54]And of course, he got the result that very low offers are not accepted. But it was foreshadowed in these psychological experiments. It was not even remarked by these people that there was something extraordinary happening. That was the reason why later ultimatum game experiments were done. This story about scientific discoveries is somewhat oblique, not very logical. Things happen in an unexpected fashion.[LXXXIII]

Chris Starmer That sounds like a very nice note to end this session on: unexpected nature of scientific inquiry. I think we will leave that hanging in the air, perhaps, over coffee and return in about half an hour. Thank you very much.

Reinhard Selten That much about methodology.

[53] **Sauermann, Heinz** ed. 1978a. *Bargaining Behavior*. Tübingen: Mohr.

[54] **Editor**: Interestingly, the Werner Güth, Rolf Schmittberger and Bernd Schwarze (1982) paper does not quote any paper from the Sauermann volume on Bargaining Behavior. Any idea why? **Reinhard Selten**: "Werner Güth was not a participant of the Winzenhohl conference. When I discussed the problem with him, I told him of the phenomenon, and then he and his research associates probably did their ultimatum game experiments before 1978 when volume 8 occurred."

# Chapter 6
# Laboratories

## The PLATO System

| | |
|---|---|
| Chris Starmer | Here we are. The final session came around much more rapidly than I had imagined. My plan is now to turn towards the final of the four themes—the labs. I wanted to explore some of the historical developments related to them. With a view of trying to pursue some chronology, I am going to begin this session by asking particular individuals about their experiences, and then, later on, draw people in. If you want to chip into those, then, feel free, but I also want to pursue a particular line. I was going to start with Stephen, actually, if I may, and ask you a bit about your first experiences of an experimental lab, which I think would probably be in Vernon's lab in Arizona. |
| Stephen Rassenti | Yeah, my first experience was in the science library at University of Arizona on the PLATO machines,[1] which were hand built by an electrical engineer of the university there. They were the first touch screens that I had ever seen, 20 years, 25 years before their time. This guy built plasma screens and they were used for PLATO, but they were public machines. There was public access, and when we would do an experiment, we would reserve a few of them, and put cardboard up beside them to prevent people seeing from one machine to the other. But also, in those early days, I did a lot of oral experiments. |
| Chris Starmer | What did PLATO do? |

[1] PLATO is an abbreviation of *Programmed Logic for Automated Teaching Operations*.



A. Svorenčík, H. Maas (eds.), *The Making of Experimental Economics*,
DOI 10.1007/978-3-319-20952-4_6

Stephen Rassenti	PLATO was a mainframe system. The mainframe was housed at the University of Illinois Champaign—Urbana, and it was a CDC [machine],[2] the biggest CDC machine that was available, back at that time, in the mid-'70s. And it was an educational system that was designed for question [and answer interaction], and it was way ahead of its time, actually. It provided us an opportunity to actually computerize some of our first experiments. It wasn't designed for that purpose. Clearly, the [so-called TUTOR] language wasn't designed for that.

Chris Starmer	So, it is a very early network allowing communication?

Stephen Rassenti	We were connected by a T1 line directly from the University of Arizona to the mainframe and we paid quite a lot of money to rent.[3]

Vernon Smith	For a dedicated, high quality line.

Stephen Rassenti	Yeah. And anybody who wanted to use PLATO had to do that. They had to rent a special phone line from Bell [Telephone] to connect with the mainframe.[LXXXIV] These were all dumb terminals and at any given time, you were sharing that mega mainframe with probably 800 users across the country. You were being swapped in and out time-wise, as you were using the machine. They would put your stuff in central memory, then, pull it out a couple milliseconds later, and put it back in, and pull it back out.

Chris Starmer	And you started summing up about what you were doing there. Can you come back to that?

Stephen Rassenti	I'm trying to recall which experiments I was doing. I think the first experiment I programmed was probably a posted offer type experiment on the PLATO system. The first experiments I did with Vernon were hand run experiments. Most of them were. Well, actually I did do something else. At the University of Arizona I used their local [mainframe] computer and [dumb] terminals that were dispersed across the campus to do my dissertation experiments.

I had a printer [and a dumb terminal] in the room that I was doing [my dissertation experiments.] These were the first combinatorial auctions.[4] People from all over campus would

---

[2] CDC or Control Data Corporation was a supercomputer firm that built the fastest computers in the world in the 1960s.

[3] **Editor**: What is a T1 line precisely? How much did it cost? **Stephen Rassenti**: "A *T1 line*, or DS1, is a digital transmission link with a capacity of 1.544 Mbps. It cost us $3,000 per month, which was a high price for data communications at that time."

[4] See Chapter 2, Footnote 76 and Chapter 3, Footnote 76.

| | |
|---|---|
| | be sending information through the mainframe. I would have to wait for a timeslot to run the optimization and then, print out the results, show everybody the results.[LXXXV] They were done using the mainframe at the University of Arizona, but in a very awkward manner. Everything was very tedious and very slow because you had to wait for service time, and then, you had to wait to have the results printed out. |
| Chris Starmer | And what had brought you to Arizona? |
| Stephen Rassenti | Well, I was there as an engineering Ph.D. student, and I took one course from Vernon. My experience in economics was very little. Actually, in that course and another course from another fellow in economics, but then, a third course from an ergonomist in the engineering department, I wrote one paper that sufficed for all three courses. I was very proud of myself. [laughter] I wrote a paper concerning in-grown preferences for anchor points in a spatial precision task. Basically, I asked people to do things on a two-dimensional screen, and allowed them to buy anchors that gave them reference points on the screen. I aggregated the preferences for these things. I was attending to a couple of different theories at one time. I did that locally. These were single subject experiments and I did that on the machine that the Department of Systems Engineering had. But then I left. I went to Bell Labs for two years. |
| Chris Starmer | When did you leave? |
| Stephen Rassenti | In '80 ... |
| Vernon Smith | You finished your thesis in '81. |
| Stephen Rassenti | Yeah, in '82 then, and I went away for two years. Then Vernon invited me back. |
| Chris Starmer | Okay, and did you stop doing experiments in those two years? |
| Stephen Rassenti | Yes. I mathematically modeled communication systems. |
| Vernon Smith | His thesis was on combinatorial auctions.[5] |
| Chris Starmer | Okay. So, that is great. |
| Vernon Smith | I have something to add to that. |
| Chris Starmer | Sure. |
| Vernon Smith | When did you come to Arizona? Seventy ... |
| Stephen Rassenti | '76. |

[5] See Chapter 2, Footnote 76.

## Technology Changes the Message Space

Vernon Smith

76, all right. We discovered PLATO in the fall of '75. I was teaching a course in experimental economics, and there wasn't that much demand for it. It was dual level [course, both] undergraduates and graduate students. A great mix. I learned a lot from the undergraduates. Mike Vannoni, an engineering undergraduate, was in that class. We were doing hand run experiments.[LXXXVI] He came to me and said, "You know, there is—over in the science laboratory a PLATO computer system. I'm familiar with it. I think this would be a great way to run experiments." I said, "Okay, well, let's go over, talk to the guy" [who ran the lab.]

PLATO was a program learning system. Most people would sit down [in front a PLATO terminal]. There would be a program for an astronomy course, a physics course, an engineering course, or a chemistry course. But it actually had provision for interactive work because the teacher could be involved. It turned out that was an incredible [feature of the] TUTOR language. This was an incredible system for doing interactive experiments, and that got us started. All *Arlington Williams* knew was a little bit of Fortran. [At that time] any person with some programming knowledge knew Fortran. Arlie was fascinated by the double auction.

And so, he set to work to create the first computerized double auction. He finished that in the summer of 1976. We ran that summer a series of 12 experiments, and we compared them with hand run [oral auction] experiments. That was the exercise. Arlie had learned so much programming by trial and error of using the TUTOR language. This [double auction] program was clumsy. He started over and completely redid it, and it came out as an incredible program. It lasted for years.

Then, there was a series of [collaborators] after that. Many of them were undergraduates like Jonathon Ketchum, but also graduate students like *Donald Coursey*, that became involved in programming, worked with me in classes, and everything. That became the primary instrument that we used. As I look back, what is fascinating is we thought that this device would make it easier to record data without making a mistake, make it easier to run experiments. It would just facilitate what we were already—had been doing for 20 years. But it totally changed the way we thought about experiments, because now you could handle really large message spaces.

You could start to do things that you wouldn't have even dreamed of doing if you were doing hand run experiments because you wouldn't be able to do it. So, very quickly, the notion of a smart computer assisted market, which is very complex for humans [without computer assistance but easy when they] are putting in decisions, and then, there is some optimization routine that processes that and then feeds it back to subjects. We never even imagined doing that with hand run experiments.

Chris Starmer I would like to come back to those issues of how the technology was perhaps changing things. I don't know whether your questions are relating to that. I would quite like to go across the Atlantic and ask perhaps Reinhard to tell us something about—you were initially doing some in Bielefeld, I think.

Reinhard Selten Initially, we did experiments from '58 to '60 for nine to ten years. In Frankfurt where we had this research group where we did the experiments, but they were mostly hand-run. We didn't have a laboratory, we just used seminar rooms. We didn't use computers. For our market experiments we used the abacus, the Chinese system. It was very difficult, but it made all sorts of computations. Computerization came much later, when I had my experiments, my laboratory in Bonn.

Chris Starmer Could you tell us something about the developments in Bonn?

Reinhard Selten We began with computerization in '84. And in the beginning, it was very difficult with three computers and we build up [the number of computers in the lab]. Finally, we had 12 computers in this very little, very narrow space for that, which meant that the people that sat at these computers and worked had to leave during an experiment. The programming was quite difficult. To program something like posted price, oligopoly or so you would need three months to program. It was quite difficult at that time. And later, in the ...

Chris Starmer But just staying with that initial lab, what were your thoughts about what a lab should be like? You are talking about the computing, but just in terms of the space and the physical set-up.

Reinhard Selten Later, I don't know when, we got better space arrangement. This is our ideal of a lab: we have cubicles with a curtain before them, clearly separated. When you see labs today where they often just have little cardboard separation, it is not a very good separation of people [subjects]. They can look at each other's screens and so on. It is not good enough. We have these terminal cubicles. At the moment there are now 24, but for a long time, we had 18.

## Austin Hoggatt's Visionary Laboratory

| | |
|---|---|
| Chris Starmer | When you were first setting up the lab, did you have any sources of inspiration for what a lab ought to be like? |
| Reinhard Selten | I was already exposed to a lab in Berkeley when I was there in '67, '68, and that is [gets interrupted]. |
| Chris Starmer | This is *Hoggatt's* lab. |
| Jim Friedman | That is right. |
| Reinhard Selten | It was in operation, and it was very good. It had six computers, six terminals. |
| Jim Friedman | More. |
| Reinhard Selten | Yeah, later it was more, but at the time I was there, there were six terminals. |
| Jim Friedman | That is when I was there. |
| Reinhard Selten | And in each—there were cubicles. In each cubicle, there was a television apparatus, and a typewriter, which was |
| Jim Friedman | Teletype terminal, a dumb terminal |
| Reinhard Selten | teletype, yeah, which go by tapes—by, now how do you call them? |
| Jim Friedman | I think it was just punch tapes. |
| Reinhard Selten | Yes, punch tapes with holes in them. For example, I did experiments there. *Austin Hoggatt* had a television camera and used it to film my instructions. Then, each subject could watch this film on the television screen in his or her cubicle. That was great. I gave the same instructions to everybody in this experiment. It was like some television show. This was a very good laboratory. It was very difficult to program there. *Hoggatt* had a big grant for putting APL on this very small computer. This was a FDP? |
| John Ledyard | [It was a] PDP-11. |
| Jim Friedman | No, no, no. [It was a] PDP-5 back at that point or shortly afterward. Just before *Hoggatt* and I did our work in that lab it was a PDP-8.[6] And the two computers were linked together. No, PDP-11 was too sophisticated. [Ledyard: Yeah] |
| Reinhard Selten | They had very small computers, and were much smaller than today personal computers. The power was very limited. [Ledyard: Much limited] But nevertheless, he succeeded in |

[6] Programmed Data Processor was the name of a series of minicomputers made by Digital Equipment Corporation.PDP-8 was the first commercially successful minicomputer and used 12-bit processors.

doing marvelous things with them. [Friedman: Yeah] It took a lot of programming, but he did this. That was already my ideal of a laboratory when I began to build my own laboratory. I think it was the first real computerized laboratory in the world.

Chris Starmer: Jim, could I just ask you about your experience?

Jim Friedman: I just really want to amplify what Reinhard is talking about Berkeley because that is the only real laboratory where I functioned. The stuff I did at Yale was done in a little psychology lab with some cubicles in it. And that was it. Or in other cases I used offices in the building where my office was located. But at Berkley, first of all, if you can imagine that lab, the space must have been two or three times the size of this room. And visualize, of course, a much lower ceiling. A two and a half to three meter high ceiling, and in the ceiling, there were tracks in the ceiling and in the floor. You could move on those tracks wall partitions that might have been two meters wide or two and a half meters wide.[7]

You could push all those things off to the edge and have a great big room. Alternatively, you could run them around and configure cubicles that were genuinely enclosed, totally enclosed rooms that were really soundproofed off from everything else. And the panels were all wired so that the terminals that Reinhard talks about and the monitors that he talks about were plugged into the electronics that were in the walls. This was probably very likely the first purpose built computerized economics experimental lab ever.

In that setup, you could run experiments. You could do something along the lines of what you were doing with the PLATO, but probably not quite as sophisticated because the computing power was lower. But you could run experiments based on models that were much too complicated to use in a hand run operation. [Smith: ehm] In fact, the experiment that Austin and I ran there had such a model. It was a model you couldn't dream of doing in a hand built operation. As a matter of fact, I think we somewhat overcomplicated the model that we used from what we ought to have done. I think that was also my fault, but that is a side issue. But anyway, the laboratory had these capabilities.

Our data, as Reinhard said, the instructions would come on the screen to each subject, and the subjects typed in choices.

[7] For a detailed description of the laboratory see: **Hoggatt, Austin C.; Joseph Esherick and John T. Wheeler.** 1969. "A Laboratory to Facilitate Computer-Controlled Behavioral Experiments." *Administrative Science Quarterly*, 14(2), 202–07.

| | |
|---|---|
| | They got responses back on the screen telling them what has happened, and all the data was being recorded. So, it is already in a form that is easily transferred onto a mainframe computer, from which you could do data analysis. |
| Reinhard Selten | In our experiments the data didn't come on the screen but by Teletype. |
| Jim Friedman | Well, maybe. I thought—now, it is a long time back, and I don't trust my memory. |
| Reinhard Selten | They were Teletyped. The computer just gave the data back on the typewriter. |
| Jim Friedman | Well, maybe that is what happened with us, too. Because, it is again, 40 years back. |
| Reinhard Selten | The screen was used for showing instructions on the television screen. |
| Jim Friedman | I think we did report back on the screen. |
| Reinhard Selten | That maybe was possible... |
| Jim Friedman | I think. |
| Jim Friedman | After all, this was—you were there '67, '68. When *Austin* and I ran our experiments, it was two or three years later, it was '71, '72, and I think the computing power was a little more at that time. |
| Reinhard Selten | Then it certainly was possible. He always improved this. He was a perfectionist. |
| Jim Friedman | That might have perfected the [experimental] technique. |
| Chris Starmer | Vernon. |

## A Laboratory Is More Than Its Physical Infrastructure

| | |
|---|---|
| Vernon Smith | I just wanted to comment that *Auggie* was a real innovator. |
| Jim Friedman | Oh yes. |
| Vernon Smith | Tremendous [one]. And I wanted to ask Jim to tell a little more about what happened to that and what the end result of that whole enterprise was. |
| Jim Friedman | Oh, you mean of our work together? Or what? |
| Vernon Smith | No, the laboratory part, what happened to it. |
| Jim Friedman | Well, I think what happened to it was that it never became an ongoing enterprise. [Smith: Yeah] It needed to be fed. It needed to have a staff, and it need to have—ideally, it would have been a place where lots of people in this room would gone to run their—would have sent stuff to do their experiments. Ideally, there would have been a staff, and you could set up your experiment, and you could have it done out |

| | |
|---|---|
| | there. That kind of thing never materialized. As a consequence, it didn't get a lot of use. I don't think *Auggie* did much experimental work, maybe no experimental work after he and I finished what we were doing. And we were done in the early '70s with our experiment.<br>Eventually, the lab was abandoned as a lab and the space was reconfigured. I think they probably tore out what was there and reconfigured it into some kind of a conventional office space or teaching space of some kind. |
| Chris Starmer | John and Vernon, you two a quick comment please. |
| John Kagel | I can't see outsourcing the experiments [Smith: ehm] to some super facility somewhere. Because for the person who is most invested, at least in my lab, in a particular experiment, sitting in there for the first two or three runs of a given experiment is absolutely critical to understanding what is going on. |
| Chris Starmer | Yeah, I think we are touching here on something I want to come to a little bit later, perhaps, sort of organization structure and how people relate to the experiments within that. So, good thing, but can I come back to it? Vernon, what did you want to add? |
| Vernon Smith | I agree. Also, Berkeley invested in infrastructure and not in people. [Selten: ehm] And people would have actually used it and be an integral part of that process. That's the thing. But it was the kind of thing that had to be learned. |
| Chris Starmer | Was *Hoggatt's* lab an influence when you were setting up your Arizona lab? Did you know that lab? |
| Vernon Smith | Oh yes, we knew. This was back in the '60s. |
| Jim Friedman | It was set up sometime in the '60s. |
| Vernon Smith | Yeah, it was back in the '60s. |
| Jim Friedman | It was there. Well, I went to Berkley for a year on leave in Fall of '66. And it already existed then. |
| Vernon Smith | It was set up in the '60s. |
| Chris Starmer | And you set up the Arizona lab in '76? |
| Vernon Smith | Yes, but before that, at Purdue, there was a lab. *Bill Starbuck* was involved in the design of that lab, primarily.[8] It was toward the end of the period that I was there.[9] *Bill* knew about the one *Auggie* had, and he knew about what had been going on around. He was involved in designing that lab [at Purdue]. Charlie [Plott] probably knows more than I do about, because he stayed on [at Purdue].[LXXXVII] That was |

[8] **Fromkin, Howard L.** Ibid."The Behavioral Science Laboratories at Purdue's Krannert School." 171–77. See also Chapter 2, Footnote 38.

[9] Smith left Purdue in 1967. Plott left Purdue for Caltech in 1971.

| | |
|---|---|
| | actually the first lab [that I was involved with]. I think it had basically a lot of the same errors. People just were not integrated with that facility. I think Bill had the idea you create that facility, and then, people will come to it. No, it just doesn't work that way. You don't farm out work. |
| John Kagel | Yeah, it is very hard to farm out work. |
| Vernon Smith | Not if you are going to be responsible for the data and for the engine for generating it. |

## Space Fights: NASA Pays for Labspace

| | |
|---|---|
| John Kagel | [About] my first computer lab. The animal lab is a totally different story. We had been doing common value auctions by hand. We knew networking ... |
| Chris Starmer | When is this computer lab? |
| John Kagel | I would just have to look at the grant [flips through his CV], which was 1984. Something like that. Yeah, 1984.[10] We had been doing the work by hand, the experiments by hand from '82 to '83, and it was just getting more and more difficult to do, especially clock auctions. We wanted a lot of speed and we knew about networking. We got a room, and we got the grant to purchase the hardware. One of us was going to do the networking. That never really worked out and we brought in—I think we had a graduate student whose husband was a networking guy, and we hired him part-time. He put the network together, and we were off and running. We knew nothing about other labs. We needed a space for the computers. Space is one of the scarcest resources at any university. [Crosstalk: Several yeahs] There can be fights over that—they will promise you space, and then, it won't be there. But then, you show them your contract. |
| Chris Starmer | Did you have to do that? |
| Vernon Smith | Oh, you don't want to go there and talk about that. [some laughter] Oh my God. |
| Betsy Hoffman | You can spend the rest of the time on space fights, you will probably want to hear from Charlie. |
| Chris Starmer | I do. Can you tell me a bit about the developments at Caltech. |
| Charlie Plott | Most of our stuff was done by hand using classrooms or anything else we could find even though the university classrooms, we would travel to other schools for subjects— |

[10] National Science Foundation: "Information Impact and Information Processing in Common Value Auctions," 7/15/84 - 7/14/89, with Dan Levin.

of course, we did not have a large subject pool at Caltech. So, we would go to PCC[11] and UCLA.

Chris Starmer Experiments by hand starting from where and for how long?

Charlie Plott The first committee experiments were done, I guess, early 1970, were done by hand. You had sheets of paper, and you had the indifference curves drawn on them, and the indifference curves had amounts of money attached to the indifference curve. We had already broken tradition. Vernon had used quasi-linear utility functions while we were out of the world [of market experiments] in public goods experiments. We were in a spatial world with big indifference curves drawn on pieces of paper. There are a lot of ways to do that. The [charts] had money attached to the indifference curves.[LXXXVIII] In a committee experiment, [subjects] then propose something and everyone else could look to see what was being proposed.

Chris Starmer Charlie, can I bring you back to tell me about the lab though?

Charlie Plott Okay.

Chris Starmer Were the experiments done in a particular space?

Charlie Plott [Experiments] were going on everywhere. Then, we had a small lab in Caltech that was a development from a speech lab.

Chris Starmer When did you have that?

Charlie Plott This would have been 1972, 1973, but it was not really outfitted appropriately. It was [designed] for committees with individuals located in soundproofed rooms. Conversation took place through a [sound control system]. Subjects could speak to each other or speak to a monitor. It was never really used for much. We used it some, but not a lot.[LXXXIX] The first real lab [at Caltech] starts in around 1984, 1985 when I got a grant from General Motors. It was a sizable grant. And we also got a grant from NASA to study the space station.[12] Now, that was a very crucial move. I got the space for a lab. *Dave Grether* was a division chairman and lo and behold [the space for a lab simply] appeared. [hits the desk]

John Ledyard He took out the art gallery...

Charlie Plott He took out the art gallery and it became the lab. [group laughter] With the space station research, which was a NASA

---

[11] PCC or Pasadena City College, is a community college conveniently located a few blocks away from Caltech.

[12] Since October 1984 Plott in collaboration with the Computer Science Department of the Southern University in Alabama was investigating various pricing mechanisms for NASA Space Station services. This motivated Plott to seek additional support from companies such as GM and IBM; private and public funding agencies. All funding was approved during summer 1987 and physical equipment was installed in the fall 1987. Plott received $400,000 from General Motors and $500,000 from the Bradley Foundation. Plott opened the *Caltech Laboratory of Experimental Economics and Political Science* (EEPS) in 1987.

project, we did two things. [We constructed the lab space and] we hired a little group of programmers in Southern University. We hired some of their faculty and one of their graduate students. The thing we needed to do was develop a local area network. There were no good computerized local area networks at the time. At that time, this was before Ethernet, we didn't even have the capacity to have simultaneous broadcast,[13] which we needed for any number of contact service.

So, we had to go back and this graduate student, who actually constructed a local area network and on top of that local area network he constructed something we call MUDA, *Multiple Unit Double Auction*, which fit on a little floppy disc. You could stick it in [in a personal computer] and bring it up and it could go for 20 markets and 20 people. Which was much larger than anything else that had existed at the time.

Chris Starmer — Do you want make some comments on MUDA?

John Kagel — No, let Charlie finish.

## Portable Laboratories

Charlie Plott — We passed this particular disc around and I would say that was a pretty major step for technology evolution in labs, because it was used at maybe a hundred different schools.

Chris Starmer — Was this the first portable, interactive software?

Vernon Smith — Yes, DOS-based.

Charlie Plott — Completely portable. Just stick the disc in the machine. It did 20 markets of double auction and it worked pretty well. We continued to use that.

Chris Starmer — Were many people porting it around?

Charlie Plott — I assume that all these guys probably saw it at one time.

John Kagel — Yeah, we used it.

Chris Starmer — When did you start using it?

John Kagel — Well, I mean we used it for a specific project.

Charlie Plott — Yeah.

John Kagel — We had a specific project and so we got the software and we used it. I figure it was at Pittsburgh at some point.

Charlie Plott — I suspect almost everyone around used it at one time or another as they were doing experiments at that time. We passed that around. Then the next installment came when I got a technology grant, an infrastructure grant from

[13] Messages sent to client terminals at the same time.

NSF.[14] We moved from MUDA, which is a local area network programmed in C. Now by this time Ethernet was in and the web had just come into being. We moved to web based programming and programming in Linux and Perl. We developed a program called *Marketscape*, which is infinitely more powerful than MUDA.

The same double auction structure. With this one, you can implement production and multiple markets. With *Marketscape* almost any economy desired can be created with production and multiple exchange rates, and multiple markets. And we studied them. The *Marketscape* program has continued to grow. Even now we are making it available on Ubuntu.[15] If you want these programs you can go to a Wiki, you can login, create a stick, then, on that stick will be my computer.[16] You can bring my computer up as a virtual machine on your computer, on any machine, and run my programs.

Now in these early years we also created a public goods program, we were interested in Vernon's public goods program. So, we were able to put that on the same ... [type of program as MUDA]

**Chris Starmer** When is that one?

**Charlie Plott** This is again when we developed MUDA software package. It also had a capacity to run public goods, which we did. We tested Vernon's idea of unanimity and public goods. Then in the late 90's we—I worked with *Charles Rivers Associates*—took a contract from the FCC. I had already developed programs for combinatorial auctions and that work for the FCC created the machinery for a new generation of smart markets which have continued to evolve and be used in many, many, many different kinds of things like electric power.

The guy who originally created the network [MUDA program] that the faculty at Southern University couldn't create, his name is Hsing-Yang Lee. I hired him, he became my senior lab technician and I have had him ever since. Then the person that we hired with the FCC program, named Travis Maroon, is an expert on dealing with the smart market aspects. I hired him and he's doing [much of the work in my lab associated with large and complex auctions and markets].

**Chris Starmer** When did those two people join the group?

---

[14] In 1995 Plott submitted a proposal for a National Science Foundation instrumentation grant. This scheme was open to all laboratory sciences and engineering on a competitive basis and grants amounted up to eight hundred thousand dollars over a two-year period. Plott became the first social scientist ever to receive such a grant.

[15] Ubuntu is a Linux-based operating system.

[16] For more details see http://marketscape.caltech.edu/wiki [Accessed on March 31, 2015].

## The Management of Laboratories, Software, and Subject Pool

Charlie Plott: Both of them have been with me for almost 20 years. And both of them are fulltime adults with families and they are highly skilled technical people. These days they are the people needed in the lab.

Chris Starmer: Yeah, I . . .

Charlie Plott: those are the people, those are . . .

Chris Starmer: I'm going to come to that in a moment, sort of the organization, structure, and the skills in the lab if I may, but before doing that could I, just, there is a second lab at Caltech, right John?

John Ledyard: Yup.

Chris Starmer: Could you perhaps tell me a little bit about the developments at hand?

John Ledyard: Well, we have a fairly large group of people who want to run experiments at Caltech. Charlie [Plott] proselytizes so well that everybody who shows up, even if they are not an experimentalist when they arrive, they want to be later. And we were having trouble getting everything done that everybody wanted to do. We basically carved out space, the rest of the art gallery [laughter] and set up a second lab which is shared by about five or six people most of the time. [17] There have been ins and outs and there is turn over. There is again, a lab manager, an individual whose fulltime job is to keep the systems running and make sure everything's going, develop, and coordinate software development sometimes.

The expenses for him are shared among the six. It is a cooperative operation. We have a director and it is supposed to rotate every two years, it hasn't always happened. People get locked in and enjoy being [a director.] It becomes temporarily that individual's lab. It is always an issue of sharing and it is an interesting problem. Most of us have a second person we hire to do software development, technical supports. Charlie has two guys. We have this one person we share, and then there is most of us have a second person that becomes our key technical development person.

But you try to get into the guts of the software yourself. You don't want to just delegate it. And it has been fairly successful. We have sort of fairly loose rules about

---

[17] The *Social Science Experimental Laboratory* (SSEL) was opened in the summer 1999 and has run by a group of Caltech experimentalists – Tom Palfrey, John Ledyard, Colin Camerer, William Zame, and formerly Richard McKelvey, Jacob Goeree, Peter Boessarts, and Liat Yariv. Palfrey and McKelvey jointly designed the laboratory.

scheduling but we don't seem to conflict with each other. We have a problem at Caltech, which is a little unique. Since we are so small we have subject pool problems. We have got 800 undergrads and if you want to run an experiment with a lot of different subjects who aren't professional lab rats you need to go someplace else so we basically set up an arrangement with the UCLA.

They found a room. They found machines. We gave them all our software so they essentially duplicated our lab at UCLA. In return for that we get access to their subject pool. It is like a second, but a remote facility where we get lots of students. That is the state of the art now.

Charlie Plott There are two more things. You see at Caltech we also have two fMRI [labs].[18] We have a group of biologists and neuroscientists that is with [economists] *Colin Camerer* and *Antonio Rangel*. They tend to work at the individual [decision] level. The second lab tends to work on game theory and game types of things. My lab tends to work on big markets, general equilibrium and these kinds of things. The fMRI labs tend to work at the individual.

By the way, regarding the subject pool problem, my research has almost entirely moved to a lab in which the subjects often do not show up in the lab. Many of my subjects are remote, I train them remotely, I recruit them remotely, and I pay them remotely. So . . .

Chris Starmer And so this is web based?

Charlie Plott It is all web based, but don't think about questionnaires. It is a lab operation, but my subjects often are all over the world. We are learning how do you maintain laboratory conditions, when your subject base is so large and so remote. That has been a major change in lab technologies, in lab procedures for me.

John Ledyard I think the character of the research done in the various labs changes over time depending on who the people are. It is not just game theory in one, markets in another. It is very much driven intellectually by the individuals who are doing experiments.

Al Roth First, Vernon mentioned when he was talking about PLATO that it had this interactive facility where the teacher and the students could interact. That was mostly used in foreign language classes. At Illinois the big cluster of machines was in the foreign language building. When we started to do experiments using that network we got permission to reserve the language lab to run experiments with. To follow up on [what was said], let me make two points. We also had a

[18] fMRI stands for functional magnetic resonance imaging.

facility that had been built for psychologists so it was rooms separated with curtained one way mirrors and things like that.

**Chris Starmer** When were you using these facilities?

**Al Roth** We probably, well we were using them at the same time probably starting in the spring of 1975. The computer technology was a little primitive. TUTOR was a difficult language to program some things in, because the computing power was not in front of the subject because it was this going back and forth. Some things were initially a little difficult like getting a clock that everyone would be seeing the same time on. If you ran them in a room, you could have a clock and everyone would see the same time. Whereas it was hard to get everyone to see exactly the same time if you wanted to cut things off, although there were ways around that.

But in terms of subject pool, that is something I think worth discussing because that has evolved a lot over the years. Initially when we wanted subjects, we posted flyers around campus and people would sign up for experiments in response to flyers and when you wanted to do another experiment, you would post more flyers.[XC] These days we have a computerized subject pool, our subjects come into our lab most often but they get recruited on the web and there is lots of database management of subjects.

You can have an experiment where you say, "I want subjects who are all college students and haven't participated in an experiment before," or you can say, "I want subjects who haven't participated in a public goods experiment," or you can say, "I want to recruit sessions that will be only men and only women."

**Chris Starmer** And at what point did you start introducing the web-based recruitment?

**Al Roth** Did we do that at Pitt or [addresses Kagel]

**John Kagel** No, we didn't have that at Pitt.

**Al Roth** So, for me it started when I came to Harvard.

**Chris Starmer** And what is the institutional structure of the lab setup now?

**Al Roth** There are two active labs, one physically located in the Harvard Business School that I'm associated with and one in the Kennedy School. The one in the Kennedy School is new and I mentioned it has got all sorts of physiological things. You can spit into test tubes and have your testosterone measured and things like that. [Some laughter] We do not, ours is a computer lab. There are breakout rooms, you can have people do something and then meet together and talk if you want and then do something again, so there are things like that.

**Chris Starmer** Who uses the lab? Who's in charge of the lab?

Al Roth — The business school is in charge of it. They have an office called something like faculty research or something that has official charge.[19] We hire a fellow each year who is the interface in the lab and is supposed to make it easy to use. The subject pool is easy to use. People all over campus use our lab. Economists use it; it is used in the business school in marketing and organizational behavior. We are pretty open. Sometimes people from the general community who don't have a lab of their own come and we try to accommodate them.

Chris Starmer — So, it is shared and you have research fellows who go there and do things. Do you go yourself to the lab very often?

Al Roth — Not for other people's experiments, the –

Chris Starmer — No, . . .

Al Roth — So, not very often, I'm often there when we are testing what the software looks like, but I no longer [conduct experiments myself]. I haven't run subjects in a while. No, almost all of my experiments are collaborations with someone. Normally, it is my coauthor who is running the subjects.

Chris Starmer — Betsy.

## Lab Funding from an Administrator's Perspective

Betsy Hoffman — I actually wanted to make some observations from the perspective of a university administrator who now runs a university that has huge investments in scientific and engineering laboratories. It is very interesting to see the development of the economics lab as [they are] beginning to look more like say an engineering lab. Maybe not a biology lab because of all the test tubes, genomic equipment and everything you have in a biology lab.

But say perhaps an engineering laboratory where you are going to have a lot of equipment. A typical engineering lab will have dedicated technicians who are on the P&S, the professional and scientific track, in other words they are not faculty. They are scientists without tenure, post docs, programmers, that depend on a very large amount of money to run them. And most of these labs now depend on a large amount of money to run them whether it is through the university as Chapman has done for Vernon, the NSF, or contract labs as Charlie [Plott] has done. Science requires a

[19] It is the Office of Faculty Research.

lot of funding and a continuous source of funding. This goes back to the [topic of] funding.

Then there is the whole space issue. One of the things we are moving to at Iowa State [University] is you have to pay for space. We just built a brand new Biorenewables [Research] Laboratory that is completely a lab building and you basically have to pay for space. In other words you have to have a grant and you have to pay to [occupy space]. If your grant runs out, you lose your lab space. I think that is what is in some sense the ultimate solution to the space problem in economics will be—paying for space.

Vernon Smith: You mean a tradable resource?

Betsy Hoffman: A tradable resource, right.

Chris Starmer: Two hands and then I will come back to you. [Betsy]

Al Roth: Just on space—an observation from somewhere else I once went to help inaugurate a new building at Case Western [University] and when there were some experimenters there, they have now all left.[20] I gave a talk of some sort and I said to them, "Let me see the new lab," and they say, "We could show it to you but it is a trolley with laptops on it. We wheel it into a classroom and we make our lab there. We don't have permanent space for a lab but we can run a lab.

Betsy Hoffman: Right. That is another solution.

## Loosing Control

Chris Starmer: As Charlie pointed out the nature of a lab is evolving in the internet age as well.

Vernon Smith: Yeah.

Betsy Hoffman: Uh-huh.

John Ledyard: Yeah.

Chris Starmer: So, I think these things are not [unfinished]

John Ledyard: There are experiments being run with a *Mechanical Turk* these days which are even a different way of doing experiments but which is

Chris Starmer: But sorry again?

John Ledyard: Mechanical, *Amazon's Mechanical Turk*.[21]

Chris Starmer: Ah, yes.

---

[20] Robert Slonim (now at the University of Sydney) and David Cooper (now at the Florida State University) were experimenters at Case Western in 1998–2004 and 1999–2007 respectively.

[21] *Amazon's Mechanical Turk* is a marketplace for hiring workers who can perform tasks from home. https://www.mturk.com/mturk/welcome [Accessed on March 31, 2015].

| | |
|---|---|
| John Ledyard | You can hire people for a very low rate. |
| Betsy Hoffman | Uh-huh. |
| John Ledyard | I mean it is still an open question to whether you got the controls you need for this ... |
| Charlie Plott | We tried that. |
| John Ledyard | What? |
| Charlie Plott | We tried that and we never had success with it. |
| John Ledyard | There are some experimental results that are exactly the same as you get in a controlled lab experiment. But it is an open question. When you talked about fund raising, I just want to emphasize that running a lab where you have a full time person in charge plus programmers is expensive. *NSF Economics Program* does not fund this kind of people. As far as I know. |
| Betsy Hoffman | Yeah. |
| John Ledyard | We have had zero luck getting these people into a budget. We have had to fund these people with a lot of different sources, but all private mostly—foundations and various other things. And it ... |
| Betsy Hoffman | I think eventually NSF is going to have to come to the realization. They fund technicians for biologists and engineers |
| John Ledyard | Sure |
| Charlie Plott | Uh-huh. |
| Betsy Hoffman | Eventually they are gonna have to come to the realization if you are gonna have serious |
| John Ledyard | One hopes. |
| Jim Friedman | One hopes. |
| John Ledyard | But I just, it's just did [sit on a NSF panel Economics and Computer Science] |
| Vernon Smith | It will never happen if the panels have to |
| John Ledyard | Right, that's right. |
| Vernon Smith | decide on it. |
| Charlie Plott | Higher level NSF has to weigh in on this. |
| John Ledyard | Yes. |
| Betsy Hoffman | Yes, that's right. That's exactly right. |
| Chris Starmer | Frans. |
| Frans v. Winden | I have a question to Al. So, you said that basically the whole campus is using one lab, right? |
| Al Roth | Well, two. I mean there is one now in the Kennedy school and one at the business school |
| Frans v. Winden | Okay. But one of these labs is for the economists as well as the psychologists? |
| Al Roth | Ah, the psychologists have their own labs. |
| Frans v. Winden | So [interrupted]. |

| | |
|---|---|
| Al Roth | No, I'm sorry. Psychologists have their own labs and they tend to be [interrupted]. |
| Frans v. Winden | Because I wondered about the rules that people have to follow. |
| Al Roth | No, so, the psychologists, each psychologist has his own lab and they tend, I think, not to be computerized. They tend more to be these rooms. For them it is a real space issue. |
| Frans v. Winden | Our university has tried to talk us into getting one lab, and then we would share it with psychologists, but you have kept these separate. |
| Al Roth | Right, no, so, I think each psychologist has their own lab at Harvard. And the economists share two labs. |
| Frans v. Winden | Right. |
| Chris Starmer | I'm going to move on to a couple of final things I wanted to ask you about which aren't really related to labs. Before I do that, I will just ask if there is anything else you wanted to say about labs? |
| Betsy Hoffman | One of the developments we are beginning to see is with everybody having their own laptop now, is the ability to set up a lab anywhere and put the software temporarily on, so the subjects actually bring their own laptops. We have someone not doing experiments, but we have someone who now wants to start teaching classes, short courses. What he was doing was commandeering like an English lab to teach this course and then he got the idea that everybody has a laptop and he said, "I can just say, 'In order to sign up for my course you have to bring a laptop." I suspect that that is going to be one of the developments that is going to come as well. |

# Chapter 7
# History and Future

## How to Do Science and How to Name a Society?

Chris Starmer: Thank you. All right, as I said there are two final things that I want to ask you guys. The first one is about disagreements. I think we have been a very jolly bunch and, I'm happy to say, nobody quarreling with one another much. But it seems to me there are some notable disagreements associated with experimental economics happening over the years and I can identify at least two different sorts of disagreements. Disagreements between experimental economists are fairly vocal, for example the debate about misbehavior in first price auctions would come to mind.

But also disagreements between folks who view themselves as experimental economists and others who identify themselves as behavioral economists. I see apparent lines of fracture between different camps using experimental methods. I was just wondering if anybody had any reflections on the nature of those disputes. Al.

Al Roth: A lot of what you think is interesting or surprising about experiments depends partly on what you expect to happen. Looking around at the world of experimenters, there are experimenters and maybe these are the ones, who mostly call themselves behavioral economists, who are looking for ways in which received theory can be shown to systematically fail, and then there are people who—looking around, Charlie [Plott] mentioned that he is always surprised when theory works, but I'm not surprised when Charlie publishes a paper in which theory works because he has published a lot of them. [laughter]

He is looking for different things in different places than *Amos Tversky* was looking when he did experiments and was



A. Svorenčík, H. Maas (eds.), *The Making of Experimental Economics*,
DOI 10.1007/978-3-319-20952-4_7

| | |
|---|---|
| | trying to draw different lessons. I would think that is one of the big things, whether you are interested in exploring—not that you can't inadvertently or in the course of your work find other things—but there are people interested in exploring how well economics and received theory works, and there are other people interested in exploring how badly it works. You can do both things. And they lead you to different directions. |
| Betsy Hoffman | To me, this is really what science is all about. One of the fractures is do we have the right name for our organization. There are some who feel very strongly about the name the *Economic Science Association* and some around this table who don't like the name the *Economic Science Association.* I happen to be in the camp that likes the name and the reason I like the name is because of exactly what Al was talking about. We are really in the business of exploring the process of science, of learning from each other, of learning from doing, of refining theory on the basis of what we learn from experiments, which to me is what science is all about. The whole idea of what the name of the organization is actually a fracture that if you went around the table I'm quite sure you could get different opinions about that. |
| Chris Starmer | I would be interested to hear one. |

## SEEing Is Believing: Armchairs on Fire

| | |
|---|---|
| John Kagel | When I was president of the ESA[1] I was thinking maybe a name change would be *Society for Experimental Economics.* So you would have SEE. Then the subtitle would be "we look for and SEE what other people conjecture." I thought that would be neat. But[XCI] [some laughter] |
| Betsy Hoffman | That is a long title. |
| John Kagel | Yes, but that would be the subtitle. It would be the *Society for Experimental Economics* and then you know there would be the subtitle that says, "We SEE and look for what other people conjecture," that would be neat. But I never took it anywhere. |
| Chris Starmer | No? |
| Al Roth | I wouldn't mind seeing the name of the society have the word experimental in it, just because that would be descriptive. [long pause] And because there are other ways of doing science, one well [interrupted]. |
| Betsy Hoffman | Well, that is true, so we could actually get into a debate about this if you would like. |

[1] Kagel was the president of the ESA from 2005 to 2007.

| | |
|---|---|
| John Kagel | Yeah. |
| Betsy Hoffman | I don't know if we... To me it was really about experiments as being one example of the scientific method and so it was also trying to get to separate us from armchair economists. But I have gotten to a point in my life that as an administrator there are fights I just do not pick anymore. It is not worth it. And this is one that is not worth picking anymore. |
| Al Roth | There is now a group of experimental philosophers and their logo is an armchair on fire. [group laughter] |
| John Kagel | Yeah, I do not think that the name is an issue worth fighting about. I think the disagreements though, in terms of [unfinished] Charlie has this view that if someone finds something unusual or that doesn't fit in his priors, he has to replicate it in his lab. I think that is great. |
| Betsy Hoffman | I do too. |
| Charlie Plott | That includes my stuff. |
| John Kagel | Well, that's fine. |
| Betsy Hoffman | Well, it should. |
| John Kagel | I think he is a little bit stronger in a sense that he says, "I just don't believe it until I replicate it." There are people who are doing stuff, who I really trust. I might have a different explanation for why they are getting what they are getting and then design an experiment around that alternative explanation. But that is a little bit different than some direct replication. I think the disputes are good because we got different [unfinished] What is the explanation for the outcomes that you see? That is where the action is. |
| Betsy Hoffman | That is. |
| Vernon Smith | Yeah. |
| John Kagel | And we are going to have different views about that. That is perfectly natural. Al, tell the story about the experimenter, I forget what area, he had a lot of good enemies... |
| Al Roth | Right. In the '95 *Handbook* we quoted a guy named von Bekesy,[2] who was a hearing researcher. But I had come |

[2] Georg von Bekesy (1960): "Another way of dealing with [experimental research] errors is to have friends who are willing to spend the time necessary to carry out a critical examination of the experimental design beforehand and the results after the experiments have been completed. An even better way is to have an enemy. An enemy is willing to devote a vast amount of time and brain power to ferreting out errors both large and small, and this without any compensation. The trouble is that really capable enemies are scarce; most of them are only ordinary. Another trouble with enemies is that they sometimes develop into friends and lose a good deal of their zeal. It was in this way that the writer lost his three best enemies." **Békésy, Georg von.** 1960. *Experiments in Hearing*. New York: McGraw-Hill. Von Bekesy (1899–1972) In 1961, he was awarded the Nobel Prize in Physiology or Medicine for his research on the mammalian hearing organ.

across his stuff and he said that the best thing you can have as an experimenter is a dedicated enemy because they will read your stuff carefully looking for the flaws. He said the trouble with enemies is sometimes they become friends and they lose enthusiasm. [group laughter]

## An Anomaly Is Just Another Regularity

| | |
|---|---|
| Vernon Smith | I think it is a good idea to be very skeptical, whether it worked or it didn't work until you have replicated it. I wouldn't put it quite so asymmetrically. Because you take a theory that is working and you have done it with pretty limited part of the parameter space where it might be tested. Theory doesn't usually tell you where to go in the parameter space to test it. It is one of the many decisions you have to [make]. And you can usually find the edges of validity and push it over. I believe that is always the case and I think that is where you get new kinds of learning when that happens. But, I think you will find plenty of anomalies without looking for them. [some laughter]<br>That is the problem with a lot of the anomalies research. It is important to emphasize both anomalies and when things work. And try to understand why they work and why they do not work. [pause] |
| Reinhard Selten | Anomalies are just as another kind of regularities and it is a matter of taste whether you speak of anomalies or regularities. |
| Vernon Smith | Yeah. |
| Reinhard Selten | I mean what you have in individual decision making is that all these anomalies are really the regularities there. It is only an anomaly from an orthodox theoretical view. But for actual behavior it is a regularity. |
| Charlie Plott | I am always struck by the natural sciences. Sometimes they use the word anomalies. But when you find a theory that is rejected, it is not really called a theory. The rejected is called a paradox. [laughter] And so they explore these paradoxes because the particle is not there that they thought that would be there. But I agree with you, they are regularities. I think that these words anomalies and other things are sometimes used in a rhetorical way that is not always that useful. But a fundamental idea that, say, Vernon and John [Kagel] articulated, everybody is articulating, is you look for the regularities, you push them around, you find out where the model starts breaking down. |

That could be a paradox or anomaly, I don't know. But the point is the theory has to be improved. I think that there is one idea here that drives me at least. I really think in terms of principles of economics. That there are principles out there of economics that hang together across these models.[XCII] They are a little loose. They tend to work. If you put together a model based on these principles, it tends to capture a lot of data.

Now, we don't really talk about that or crystallize them or make them explicit unless it is a Nash equilibrium, a competitive equilibrium, or something like that. But they are out there and that is basically what we are about trying to find—real articulated basic themes that will capture as many of these anomalies, regularities we can [unfinished]. I think we are all in under the same roof.

## Relics of the Past: And of the Future

| | |
|---|---|
| Chris Starmer | On that unifying note let me come to the last thing that I want to ask you. I need to ask Charlie something, in fact, before I ask the rest. At Caltech, I seem to remember, you had a glass case that you are populating with |
| Charlie Plott | My museum? |
| Chris Starmer | A museum. |
| John Ledyard | Relics. |
| Chris Starmer | Can you tell me just a little bit about your museum? Because that is going to lead to a quick question that I then have for the group. |
| Charlie Plott | What is in my museum? |
| Chris Starmer | Yeah. |
| Charlie Plott | Well, actually it is just one of the first little stand alone desktop computers that we used. This was even before the [unfinished], I have even forgot what it was now before an AT, the, that we used. |
| Betsy Hoffman | Was it the 8086? |
| Charlie Plott | Or was it the XT? Maybe it was the 8086, could be just a little one. |
| Chris Starmer | When did you decide to start your museum? |
| Charlie Plott | When we got some new computers. We were cleaning things up, I started throwing things away, and I thought, "You know, some of this stuff is really interesting." |

| | |
|---|---|
| Chris Starmer | When do you think that was, when did you start throwing things away? |
| Charlie Plott | Oh, I don't know, let's see, it has been, mid '90's. [Smith: Yeah] And started collecting things—the way that we induced preferences in committees, when we ran the FCC auctions by hand even though they were supposed to be computers. I kept the little workbooks that we had that were important for the backup system that was used in the FCC auctions. |
| Chris Starmer | And you have got a bingo cage? |
| Charlie Plott | I have got a bingo cage. That used to be our random variable. |
| Chris Starmer | As I understand it you are collecting this set of no-longer-used experimental tools. Here is my final question to anybody who would like to contribute. I am interested to know your predictions or preferences over what should be the next additions to the museum of no-longer-used experimental tools. [some laughter] |
| Jim Friedman | I have a set of three icosahedron random number generating dice. [some laughter] |
| John Ledyard | Dungeons and Dragons got us a long way.[3] |
| Chris Starmer | Charlie's probably accepting donations, probably so. |
| Charlie Plott | Yeah, I take up donations. |
| Chris Starmer | Any more suggestions? |
| Charlie Plott | I think that wired computers are no longer going to be available. Everything is going to be wireless after while. That is likely to happen. It could be your laptop is going to be on the way out. You are going to find experiments over your telephone |
| Betsy Hoffman | Or I think the iPad. |
| Charlie Plott | iPad, yeah. |
| Betsy Hoffman | The little ones [smartphones] are really too small to see. But I bet you in a short bit of time everybody's going to have an iPad type device that is big enough so you can actually see things |
| Charlie Plott | Uh-huh. |
| Betsy Hoffman | but so light that it carries around like a book. |
| Stephen Rassenti | That maybe your problem, but *Bart Wilson* goes around with a bunch of these little things and manages just fine with high schoolers and young college students. |
| Vernon Smith | He carries them in his briefcase. |
| Stephen Rassenti | Yeah. |

[3] Reference to a board game Dungeons and Dragons, where the dungeon master owns a collection of dices. Rolling a dice can be used to play a mixed strategy in experimental games.

| | |
|---|---|
| Betsy Hoffman | And maybe that is it, maybe I'm just too old to read things on this… |
| Stephen Rassenti | Well, they are bigger than that, but you know, they are three inches by four inches. |
| Betsy Hoffman | But they are bigger than that? But I think it is what people are going to have themselves. Because I actually think we are going to get to the point where it is, people are going to bring their own tools. |
| Stephen Rassenti | Yeah, but for the sake of control, the whole lab comes in one little suitcase and weighs a couple of pounds, you know, so it is really not necessary to demand that they bring their own, you can just pass them out wherever you are. |
| Betsy Hoffman | Yeah, I have worked with him on that. |
| Charlie Plott | Part of the answer is where you think that experiment is going to go. Obviously I think that large scale, interdependent systems are to me where part of the future is. Other people are going to say you have got to go deeper into the mind. That is going to require a completely different construction of a lab. Are you going to do chemistry and physiology of decision-making? Now that is going to be a different kind of a lab. I think that it is probably almost certainly we are going be moving more into neuroscience. Which is going to change a lot.[xciii] |
| Betsy Hoffman | That is right. That is how you will tap into real money… |
| Stephen Rassenti | On that note, I would say that you are way too quick to put your bingo cage back in …[some laughter] |
| Charlie Plott | It is still in my museum. |
| Stephen Rassenti | And I would not give away your dice either. |
| Chris Starmer | I agree. I'm close to the end of everything. Before I finally finish though, I would like to say very much thank you to *Harro* [*Maas*] and *Andrej* [*Svorenčík*] for organizing this. And thanks to all of you very much for agreeing to come and participate in it. Perhaps *Harro* wants to say something, too. I don't know. But from my point of view, I was surprised, to be honest, to [be asked] do this, and rather nervous about doing it because this is a completely new experience for me. *Andrej* and *Harro* have been very patient in talking me through this and coaching me. But I must say, it has been an enormously interesting experience. And one I feel privileged to have taken part in. Just from me, a personal thanks very much to all of you. [applause]. |
| Harro Maas | I do, of course, want to thank, first of all, you all, that you have been willing to make this tremendous long trip for just two days to Amsterdam. I found it an extremely interesting |

experience. I hope a rewarding experience for you as well. We will talk about that a bit at lunch. I would also like to thank *Andreas* [*Ortmann*] for having written this first piece that we sent you. It was at some point quite a help in getting you started. Then, Chris, you did a fantastic job in moderating the seminar, and I think it is worth an applause. [applause]

And then, the last to thank, of course, is *Andrej*, because *Andrej* did so much to prepare this witness seminar, in taking the interviews with you and in thinking about the things that should be covered at the witness seminar itself. Of course, things always turn out a bit different than you think, but I think most of it, we can be very, very pleased with. *Andrej* it was great that you prepared all of this [applause]. Thank you very much. There is a last thing. We go outside to the next room and we will gather for the photograph. And then, we will go to *De Waag* for lunch and a farewell drink.

# Chapter 8
# Biographies of Participants

**Fig. 8.1** Group photograph From left to right: James [Jim] Friedman; Stephen Rassenti; Frans van Winden; Alvin [Al] Roth; Charles [Charlie] Holt; John Kagel; Vernon Smith; John Ledyard; Charles [Charlie] Plott; Elizabeth [Betsy] Hoffman; Reinhard Selten

 
A. Svorenčík, H. Maas (eds.), *The Making of Experimental Economics*,
DOI 10.1007/978-3-319-20952-4_8

**James [Jim] W. Friedman**

| | |
|---|---|
| Born | 1936 |
| Current affiliation | Emeritus Professor of Economics, University of North Carolina at Chapel Hill |
| Education | 1960–1963 Ph.D. Economics, Yale University |
| Dissertation | The Theory of Oligopoly (supervised by William Fellner) |
| Affiliations | Yale 1963–8, Rochester 1968–83, Virginia Polytechnic 1983–5, University of North Carolina at Chapel Hill 1985–present |

On a suggestion of Martin Shubik, Jim Friedman became an early adopter of the experimental method. His dissertation work was funded by the Cowles Foundation where he remained after graduation. Friedman spent the academic year 1966–7 at Berkeley where he interacted with John Harsanyi and Reinhard Selten. Towards the end of his stay he met Austin Hoggatt, with whom he conducted oligopoly experiments, using Hoggatt's experimental laboratory. By the end of that academic year he made the conscious decision to leave experimental research and to focus on theory instead. The decision to abandon experimentation was based partly on his early desire to become a theorist. For Friedman experiments were a detour from his original plan. Nevertheless, he successfully applied with Austin Hoggatt for an NSF grant "Theoretical Research and Collaborative Experimental Research on Micro-Economic Games" (1969–71). The grant had not only a game theoretic part, but also an experimental one. The evaluation report claimed that the experimental part prompted NSF to award the grant. A decade later, the experimental results were published as a stand alone volume in a series edited by Vernon Smith, *Research in Experimental Economics*. His experimental work was largely ignored, but his theoretic research has received by far the widest attention in the course of Friedman's career. He served as associate editor of *Econometrica* from 1975–81. Friedman attended the game theory year organized by Reinhard Selten at the Bielefeld Center for Interdisciplinary Research in 1987–8 and was a regular attendant of the experimental economics meetings in Tucson in the late 1970s and 1980s.

**Elizabeth [Betsy] Hoffman**

| | |
|---|---|
| Born | 1946 |
| Current affiliation | University of Iowa |
| Education | 1974–1979 Ph.D. Economics, Caltech & 1972 Ph.D. in History, University of Pennsylvania |
| Dissertation | Essays in Optimal Resource Allocation Under Uncertainty with Capacity Constraints (supervised by Roger Noll) |
| Affiliations | Northwestern 1978–1981, Purdue 1981–1986, University of Wyoming 1986–1988, University of Arizona 1988–1993, Iowa State University, 1993–1997, University of Illinois at Chicago 1997–2000, University of Colorado System 2000–2006, Iowa State University 2007–present |

Betsy Hoffman was first trained as an economic historian in its then rapidly emerging new cliometric approach. She came to experimental economics during her second doctorate at the advise of John Ferejohn, graduate students advisor at Caltech, who suggested that no one should leave Caltech "without taking Charlie Plott's experimental economics class," which he considered the "hottest thing" in town. She took the class and some of her earliest publications were developed in it. Later on, when she was on the job market, Ferejohn advised her not to talk about her experimental economics work. Rather she should focus on her dissertation in a general equilibrium framework that dealt with the water allocation in the Colorado River Compact. After getting her first job at Northwestern she conducted a series of experiments examining various aspects of the Coase Theorem with her classmate from Caltech, Matt Spitzer, who specialized in Law. This eventually led to the award of the first Ronald Coase Prize for excellence in the study of law and economics. In the course of the 1990s she took more and more administrative roles and served as Associate Dean at the University of Arizona, Dean at Iowa State University, Provost of the University of Illinois at Chicago as well as Iowa State University, and President of the University of Colorado System. Hoffman served for three years on NSF's Economics Review Panel (1989–1991) and was a National Science Board member from 2000 until 2008. She was the fourth president of Economic Science Association between 1989 and 1991.

**Charles [Charlie] A. Holt**

| | |
|---|---|
| Born | 1948 |
| Current affiliation | University of Virginia |
| Education | 1973–1977 Ph.D. Economics, Carnegie-Mellon University |
| Dissertation | Bidding for contracts (supervised by Morris H. DeGroot) |
| Affiliations | Minnesota 1976–83, Virginia 1983–present |

During graduate school at Carnegie Mellon, Charlie Holt had no exposure to experimentation, but to some extent to Herbert Simon's behavioral economics. His dissertation dealt with auctions with incentive parameters in contracts in procurement auctions using Vickrey's game-theoretic approach. Once his dissertation chapters appeared in AER and JPE, he began receiving referee requests for papers dealing with auctions. Around 1979–80 one of them was a paper by Vernon Smith on posted offer and sealed bid auctions. He liked it and thought of it as very clever. He started thinking about doing experiments as he saw that they could provide relevant data for his theoretical work which could not be obtained through other sources. Holt's first NSF proposal on Signaling Auction Markets got funded (1980–82), although the panel review suggested doing some experiments. The second proposal he submitted was also theoretical, but it got rejected. By then he had conducted some experiments and decided to submit a purely experimental proposal which got accepted and since then he has been submitting experimental proposals. Holt learnt how to do experiments partially through trial and error, but also through attending conferences such as the Tucson meetings.

When the University of Virginia hired Holt in 1983, he was not seen as an experimentalist but as an industrial organization economist. Together with Arthur Schram from CREED in Amsterdam, he was the founding co-editor of Experimental Economics, a journal published by the Economic Science Association. He actively promotes the experimental method, by means of his co-edited 1993 textbook (with Douglas Davis) and web-based interactive games and markets for teaching and research. He was ESA's fifth president from 1991 to 1993.

**John H. Kagel**

| | |
|---|---|
| Born | 1942 |
| Current affiliation | Ohio State University |
| Education | 1966–1970 Ph.D. Economics, Purdue University |
| Dissertation | Factor Demand Functions For Labor And Other Inputs By Northwest Wheat-Fallow Farms (supervised by Robert Basmann) |
| Affiliations | Texas A&M 1969–82, University of Houston 1982–1988, Pittsburgh 1988–1999, Ohio State University 1999–present |

John Kagel was a graduate student at Purdue from 1967 until 1970. During these years he remained unaware of the experimental economics course that was taught there until the mid-1960s. With his classmate, Raymond Battalio, Kagel was supervised by Robert Basmann, an econometrician who was interested in testing economic theory against better controlled data than field data allowed for. After graduating, Kagel followed Basmann to the University Texas A&M University. Battalio joined them there as well. In 1970–71 they learnt about token economies and their use for therapeutic purposes. These economies promised the kind of rigorous data Basmann had searched for. The idea of token economies was that human behavior could be modified with the appropriate stimuli, so-called positive and negative reinforcements. Around 1973 Kagel and Battalio, in collaboration with operant psychologist Howard Rachlin and his student Leonard Green, started running experiments with animals in operant chambers. In a series of articles they also demonstrated the existence of (subject-specific) Giffen goods in such a setting; they studied the income and substitution effect; and the fanning out of indifference curves, to name a few.

Kagel left Texas A&M for the University of Houston in 1982. Kagel and Battalio continued working together, having a joint NSF grant running until 1991. In Houston, he began a long-term experimental and theoretical collaboration with a young MIT-educated theorist Dan Levin, mainly on auction theory, the winner's curse in particular. In 1988 Kagel was hired by Pittsburg with the promise of being able to build a computer lab there. He joined Al Roth who in the second half of the 1980s made a definite commitment to including experimental economics in his research portfolio. He co-edited, with Al Roth, the very successful 1995 *Handbook of Experimental*

*Economics*. He served as the president of the Economic Science Association from 2005 until 2007.

**John O. Ledyard**

| | |
|---|---|
| Born | 1940 |
| Current affiliation | California Institute of Technology |
| Education | 1963–67 Ph.D. Economics, Purdue University |
| Dissertation | A Convergent Pareto-Satisfactory Non-Tâtonnement Adjustment Process For A Class Of Unselfish Exchange Environments (supervised by Stanley Reiter) |
| Affiliations | Carnegie-Mellon 1967–1970, Northwestern 1970–1985, Caltech 1985–present [visitor 1977–8,1983-4] |

During his graduate years at Purdue, John Ledyard followed several courses with Vernon Smith, but never encountered or heard about economic experiments. During his years at Carnegie Mellon, simulations and experimental games research was around, though he did not participate in it. He spent the period 1970–85 at Northwestern where research on mechanism design and game theory with asymmetric information was spearheaded. Ledyard originally wanted to become a physicist or an engineer, which attracted him to mechanism design and later to experiments. Around 1976–7 Smith visited Northwestern, and when talking about the so-called Groves-Ledyard mechanism, Smith suggested to test it experimentally, which eventually had substantial impact on Ledyard's decision to take an active interest in experimental economics. He started to attend the Tucson meetings in the late 1970s, though still not actively pursuing experimental research himself. That changed slowly in the course of the 1980s. For instance, Ledyard consulted NASA's Jet Propulsion Laboratory about allocation of space on its future space station, which included experimental tests of the proposed allocation mechanism. Though Ledyard still perceives himself as an outsider to the experimental economics community, it is fair to say he slowly turned from a passive observer, with a main interest in theory, into an experimentalist. Ledyard served as president of the Public Choice Society, 1980–82, and on the Advisory Panel on Economics of the National Science Foundation 1978–80.

**Charles [Charlie] R. Plott**

| | |
|---|---|
| Born | 1938 |
| Current affiliation | Caltech |
| Education | 1962–5 Ph.D. Economics, University of Virginia |
| Dissertation | Influence of decision processes on urban renewal (supervised by Jim Buchanan and Ronald Coase) |
| Affiliations | Purdue 1965–1970, Caltech 1971–present |

Charlie Plott's interest in the broad and overlapping nature of economics, public economics, public choice and political science was nurtured first during his graduate studies at the University of Virginia and later at Purdue, where he overlapped for two years with Vernon Smith. Conversations with Smith on the Edgeworth box in the early 1970s made him seriously consider experimentation while pursuing work in axiomatic social choice theory. In 1973–74 he invited Vernon Smith for a year to be a Fairchild scholar and together they offered a seminar on experimental methods that Plott has taught at California Institute of Technology ever since. Since then he has ran hands-on workshop style seminar on experimental methods in economics and political science for undergraduate and graduate students. A number of its attendees went on to become successful experimental economists such as Betsy Hoffman, Mark Isaac, Ross Miller, Colin Camerer, Shyam Sunder, Charles Noussair, and Yan Chen. Plott was among the first experimentalists to study voting, agenda setting, public goods, asset markets, rational expectations and general equilibrium to name a few. In 1987 Plott founded the Laboratory for Experimental Economics and Political Science (EEPS) which served as a model for others around the world. Plott's many-sided development of the experimental method (with notions as test-bedding, complex many market auctions controlling experimental subjects at a distance) became an important factor in the spread of the experimental method in the economics community. From his first contract research on the dry barge market in the mid-1970s, he combined applied policy and experimental work. Plott was closely involved in the establishment of the Economic Science Association and its journal. He has served, amongst others, as president of the Public Choice Society (1976–1978) and as the second president of the Economic Science Association (1987–1988).

**Steven J. Rassenti**

| | |
|---|---|
| Born | 1949 |
| Current affiliation | Chapman University |
| Education | 1981 Ph.D. Systems Engineering, University of Arizona |
| Dissertation | Zero/one decision problems with multiple resource constraints: algorithms and applications (supervised by Robert Bulfin and Vernon Smith) |
| Affiliations | Bell Laboratories 1982–84, University of Arizona 1984–2001, George Mason University 2002–2008, Chapman University 2008–present |

After his BA in mathematics, Steve Rassenti looked at various fields of engineering for graduate school, especially in system engineering. He ended up in Arizona where he choose economics for his required minor. One of its courses was on experimental economics which was taught by Vernon Smith. It appealed to him greatly, because of his interest in institutional design, although Smith's focus at the time was rather on testing theories in simple environments. In his dissertation. Rassenti developed an algorithm for solving a new type of allocation mechanism, so-called combinatorial auctions. He applied it to the allocation of takeoff and landing slots to competing airlines and to public broadcasting. After graduating, Rassenti spent two years at Bell Laboratories before returning to Arizona in 1985, where he joined Smith's Economic Science Laboratory as a research scientist. He focused on conducting experiments on networked type goods such as natural gas and electric power where his programming expertise was valuable. Much of his work has been of applied research nature. Rassenti has also been concerned with the infrastructure of a well-designed laboratory. When Smith moved to George Mason University, Rassenti designed a new laboratory there. He was equally involved in the design of laboratories at Chapman.

**Alvin [AL] E. Roth**

| | |
|---|---|
| Born | 1951 |
| Current affiliation | Stanford |
| Education | 1971–4 Ph.D. Operations Research, Stanford |
| Dissertation | Topics in Cooperative Game Theory (supervised by Robert B. Wilson) |
| Affiliations | University of Illinois 1975–82, Pittsburgh 1982–1998, Harvard 1998–2012, Stanford, 2012–present |

Al Roth entered Columbia's Engineering school at the young age of 16 where he soon became interested in game theory that he also pursued during his graduate studies at Stanford. His first job was at the University of Illinois at Urbana-Champain. One of his colleagues in the business school was Keith Murnighan, who got his Ph.D. in Social Psychology at Purdue and knew how to do experiments. They decided to join forces which led to a dozen papers primarily experimentally investigating bargaining. They initially submitted papers to psychology and management science journals. Roth's idea was that psychologists would perform experiments on the theories developed by game theorists, only to find out that both communities followed very different research agendas, necessitating game theorists to do the experiments of their interest themselves. Roth consistently forged alliances with collaborators with strong experimental credentials such as Keith Murnighan and John Kagel. Almost despite his own strong bend to theory, Roth had a keen awareness of practical problems with game theoretic characteristics, such as the matching market for American medical doctors, then called the National Intern and Resident Matching Program, which led to the work he is best known for, the design of matching markets for physicians, the high school matching system in New York and Boston, and organ donations (e.g. kidneys); work that built on earlier theoretical work of Llyod S. Shapley with whom he shared the Nobel memorial prize in economics in 2012. With John Kagel, Roth edited the *Handbook of Experimental Economics* (1995), which served the explicit goal to lower the barriers to entry to the field. He was not actively involved in the establishment of ESA, but served as its president from 2011–2013.

**Reinhard Selten**

| | |
|---|---|
| Born | 1930 |
| Current affiliation | University of Bonn |
| Education | 1957–61 Ph.D. Mathematics, Johann Wolfgang Goethe-Universität, Frankfurt am Main |
| Dissertation | Evaluation of n-person Games (supervised by Ewald Burger and Wolfgang Franz) |
| Affiliations | Frankfurt am Main 1957–67, Free University of Berlin 1969–1972, Bielefeld 1972–1984, Bonn 1984–present |

Reinhard Selten's youth was marked by the Nazi-period and Second World War. While studying mathematics (and physics) in Frankfurt am Main, Selten turned to extensive form game theory on which he wrote his Master and Ph.D, theses. During these years, he also took courses in (experimental) psychology that opened his eyes for the possibility of the experimental method in economics. In 1957, he was hired by the economist Heinz Sauermann, and remained in Frankfurt throughout the 1960s heading a small group of experimental economists. Sauermann was an important organizing force for the German economic community and a strong promoter of the experimental method. Selten's first experimental work is in oligopoly theory jointly written with Sauermann. His reading of Herbert Simon's work on bounded rationality during his graduate years made him critical toward standard economic theory, which he only considered useful as a normative benchmark. Via Sauermann, Selten also became well acquainted with Oskar Morgenstern who introduced him to American game theorists and mathematical economists.

During the 1960s and 1970s he collaborated with John Harsanyi on problems of incomplete information and strategic bargaining for the military. In these years, he also met Austin Hoggatt, whose fully computerized laboratory was a source of inspiration for his own lab when he moving from Bielefeld to Bonn in 1984. Amongst others with Reinhard Tietz and Sauermann, Selten was one of the founders of the German Society for Experimental Economics Research. Selten received the Nobel memorial prize in economics for the introduction of the concepts of sub-game perfect and perfect equilibria in economics jointly with Harsanyi and Nash in 1994.

**Vernon L. Smith**

| | |
|---|---|
| Born | 1927 |
| Current affiliation | University of Virginia |
| Education | 1955 Ph.D. Economics, Harvard University |
| Dissertation | A theoretical and empirical inquiry into the economic replacement of capital equipment (supervised by Wassily Leontief) |

| | |
|---|---|
| Affiliation | Purdue 1955–67, Brown 1967–8, University of Massachusetts 1968–75, University of Arizona 1975–2001, George Mason University 2001–2008, Chapman University 2008–present |

Vernon Smith grew up in Kansas in the days of the Great Depression. After studying electrical engineering at Caltech and economics at the University of Kansas, he enrolled in Harvard's Ph.D program, where he attended, amongst others, Chamberlin's class on Monopolistic Competition, in 1952. Chamberlin used classroom experiments to show the deficiencies of the competitive market price model. During his early years at Purdue, Smith used similar classroom experiments, but by allowing repeated trading, he found support for the model. This led to his first experimental publication that appeared in 1962. In 1961 he briefly met the psychologist Sidney Siegel at Stanford, whose approach to experimentation became an important influence. Smith taught a graduate course in experimental economics from 1963–67. In this period he started developing his theory of induced valuation. By the end of the 1960s and beginning of the 1970s, Smith took a break from active research and teaching of experimental methods. He returned to them when he visited Caltech as a Fairchild Scholar in 1973–74. Jointly with Charlie Plott, he taught a course on experimental methods that had many Caltech faculty amongst its attendees such as John Ferejohn, Morris Fiorina, Roger Noll, James Quirk, and William Riker who was also a Fairchild scholar that year. In 1975 Smith moved to Arizona where he remained until 2001. These years were key to a number of experimental efforts, with students such as Arlington (Ted) Williams, who attended Smith to the possibilities of the PLATO system and was to program double auctions, and Steve Rassenti, who would program it for combinatorial auctions. Smith also engaged in a more than a decade long experimental and theoretical research program on auctions with James Cox and James Walker. In 1985, Smith and his associates such as James Cox, Mark Isaac and Kevin McCabe founded the Economic Science Laboratory. In the second half of the 1980s Smith's research group delved into contract research on utilities (electric power industry), which fitted into the world-wide efforts on deregulation, privatization and liberalization. Smith was the first president of the Economic Science Association in 1986. He was awarded the Nobel Prize in Economic Sciences in 2002 for his groundbreaking work in experimental economics.

**Chris Starmer**

| | |
|---|---|
| Born | 1961 |
| Current affiliation | University of Nottingham |
| Dissertation | (1992) Exploring the theory of choice under uncertainty by experimental methods (supervised by Robert Sugden) |
| Affiliations | 1992–1999 University of East Anglia, 2000–present University of Nottingham |

Chris Starmer completed his undergraduate education at City of Birmingham Polytechnic after which he studied for his master's degree at the University of East Anglia. He continued at East Anglia for a Ph.D with Robert Sugden where he stayed as a (senior) lecturer until his appointment as full professor at the University of Nottingham in 2000. At the University of Nottingham he co-taught for a few years courses with the late historian of economic thought, Bob Coats. Starmer's main research interests are in individual and strategic decision-making, experimental economics and the methodology of economics. He is a co-author of *Rethinking the Rules* (Nick Bardsley, Robin Cubitt, Graham Looms, Peter Moffatt, Chris Starmer and Robert Sugden, 2010), and has been on the editorial board of *Experimental Economics*.

**Frans A.A.M. van Winden**

| | |
|---|---|
| Born | 1946 |
| Current affiliation | University of Amsterdam |
| Education | 1981 Ph.D. Economics, Leiden University |
| Dissertation | A Two-Sector, Two-Department Model of the interaction between State and Private Sector (supervised by Bernard van Praag) |
| Affiliations | Leiden 1973–1980, Utrecht 1980–83, Amsterdam 1983–present |

After majoring in economics at the University of Amsterdam, Van Winden became a PhD student and eventually assistant professor in Leiden. His dissertation focused on lobbying of interest groups and their influence on government policy. He learned game theory by self-study as there was hardly academic interest in it at the time in the Netherlands. Classes in social psychology and a meeting with Selten while working on his thesis were important markers, though he was not aware of economic experiments in this period. In 1983 he became full professor at the University of Amsterdam where he started to get in touch with experimental economists through the Public Choice Society. With his former Ph.D student Arthur Schram, he was awarded a major grant that enabled him to establish the Center for Research in Experimental Economics and political Decision Making, CREED. The grant was awarded on the explicit condition to involve (experimental) psychologists as well as to disseminate the results and the method of the experiment within the Netherlands, for which they decided to run a series of annual workshops that enlisted American and European participants. From his interest in public choice, Van Winden moved to the experimental study of the role of emotions in economic decision making, branching out to experiments in (neuro)physiology. Frans van Winden served as president of the European Public Choice Society in 1986–1987.

# Chapter 9
# Episodes from the Early History of Experimentation in Economics

**Andreas Ortmann**

## Introduction

The present article builds on a background paper that was commissioned for a "witness seminar" in 2010 that had a dozen prominent experimental economists—witnesses, indeed—discuss the origin and evolution of experimental economics. Rather than providing a history of the experimental method in the behavioral science (with particular emphasis on those practices that informed experimental practices in economics), I was asked to provide exhibits from the early years of experimental economics. I was asked to refrain from interpretation and evaluation: "any suggestion of a linear history (as for example when, how and why the experimental methods in economics departed from those in psychology) should be avoided. . . . it is absolutely crucial to the success of the Witness Seminar to have the paper written open-ended, highlighting the questions at the time about specific episodes in experimenting. . . ." (email from Harro Maas and Andrej Svorenčík12/12/2009).

The five episodes that I present below are meant to be "museum pieces"; their purpose was to trigger memories and initiate discussions. For each museum piece, I sketched its context, then I summarized it (making occasionally, and quite intentionally, heavy use of quotations), and then I highlighted the methodological questions that the particular episode illustrated.

I dedicate this manuscript to Raymond C. Battalio who surely would have been among the eminent invitees to the witness seminar had he not passed away, at age 66, Dec. 1, 2004. He was an audacious pioneer of experimental economics. He was also a wonderful teacher and mentor. I am grateful for all he taught me.

A. Ortmann (✉)
School of Economics, Australian School of Business, University of New South Wales, Sydney, NSW 2052, Australia
e-mail: a.ortmann@yahoo.com



A. Svorenčík, H. Maas (eds.), *The Making of Experimental Economics*,
DOI 10.1007/978-3-319-20952-4_9

## Episode One: The Wallis–Friedman (1942) Critique of the Thurstone (1931) Experiment

### Context

The Wallis–Friedman critique of Thurstone's (1931) study about the experimental constructability of indifference maps has received prominent play in several places (e.g., MacCrimmon & Toda, 1969; Castro & Weingarten, 1970; Kagel, 1972; Battalio et al., 1973; Kagel & Battalio, 1980, Roth, 1995; Moscati, 2007; Lenfant, 2009).[1]

### Summary

Thurstone was a professor of psychology at the University of Chicago, and his experimental study was motivated through "numerous conversations about psychophysics with my friend Professor Henry Schultz" (Thurstone, 1931, p. 139). Schultz's major interest was the measurement of utility and demand functions (Schultz, 1933, 1938). Thurstone—acknowledging his own limited knowledge of economics—credits Schultz with the problem formulation and the suggestion to apply the experimental method to this problem in economic theory (Thurstone, 1931, p. 139).

Lenfant (2009), based on considerable sleuthing, suggests how the Thurstone study fit into Schultz's overall research agenda and how Wallis and Friedman got into the game. The bottom-line is they met as graduate students of Schultz in 1934 and eventually overlapped for years at a time at the National Resources Committee and the Statistical Research Group (Lenfant, 2009, p. 19). Lenfant conjectures that Friedman's interest in the Thurstone study may have been the result of his significant contribution to Chaps. 18 and 19 of Schultz (1938) and accompanying discussions that Schultz and Friedman must have had in this context. Be that as it may, Wallis and Friedman (1942) ended up as a joint contribution to a volume in memory of Henry Schultz. The alphabetical inversion of names on the chapter is unusual and probably reflects the relative contributions; it is in this context noteworthy that Wallis graduated in psychology and economics from Chicago and Columbia (Lenfant, 2009, p. 19).

Thurstone (1931) tried to trace out through questionnaires the indifference maps for hats, shoes, and coats of a group of girls.[2] Specifically, he offered his subjects,

[1] Interestingly, Lenfant documents that three reviews of Thurstone's article that preceded the Wallis–Friedman chapter gained little traction.

[2] In his article, Thurstone presented only the data of one of his subjects stressing that "the same procedures have been tried on several other subjects with similar results" (Thurstone, L. L. 1931. The Indifference Function. *Journal of Social Psychology*, 2, 139–67).

hypothetically, various bundles of commodities (e.g., hats vs. pairs of shoes and hats vs. overcoats) and then constructed from their responses indifference curves. He even estimated parameters that he used for out-of-sample prediction (e.g., the subjective trade-off between shoes and overcoats).

Although Thurstone, a psychologist, cared about subtleties such as experimenter expectancy effects (e.g., Rosenthal & Rubin, 1978; Rosenthal, 1994; Ortmann, 2005; Zizzo, 2010), Wallis and Friedman (1942) critiqued his experiment on several grounds, foreshadowing what nowadays is often referred to as the artificiality critique (e.g., Schram, 2005). To wit:

> For a satisfactory experiment it is essential that the subject gives actual reactions to actual stimuli. ... Questionnaires or other devices based on conjectural responses to hypothetical stimuli do not satisfy this requirement. The responses are valueless because the subject cannot know how he would react. The reactions of people to variations in economic stimuli work themselves out through a process of successive approximation over a period of time. The initial response indicates only the first step in a trial-and error-adjustment. (Wallis & Friedman, 1942, pp. 179–80)
>
> If a realistic experimental situation were devised, it would, consequently be necessary to wait a considerable time after the initial application of the stimulus before recording the reaction. Even an experiment of restricted scope would have to continue for so long a period that it would be exceedingly difficult to keep 'other things the same'. (Wallis & Friedman, 1942, p. 180)

Wallis and Friedman proposed an experiment in which children, day after day, would be offered various combinations of candy and ice cream, of which they each would choose one to consume. Wallis and Friedman had several objections to their own proposal (e.g., related to the stability of preferences across time and preferences for variety) and concluded that "it is probably not possible to design a satisfactory experiment for deriving indifference curves from economic stimuli" be it for the simple reason that it would be difficult to keep other things the same (Wallis & Friedman, 1942, p. 181).[3]

[3] Later authors, namely, Rousseas, S. W. and Hart, A. G. 1951. Experimental Verification of a Composite Indifference Map. *The Journal of Political Economy*, 59, 288–318, Maccrimmon, K. R. & Toda, M. 1969. The Experimental Determination of Indifference Curves. *The Review of Economic Studies*, 36, 433–451, Moscati, I. 2007. Early Experiments in Consumer Demand Theory (1930–1970. *History of Political Economy*, 39, 359–402) begged to differ. Moscati documents that the Rousseas and Hart article, save a couple of citations, had little traction (ibid.). The MacCrimmon and Toda article did better, as it managed to attract in the following two decades about a dozen citations in the journals of economics, business, and finance listed in JSTOR.

The extended discussion of methodological issues that might mar an experiment's validity, Maccrimmon, K. R. and Toda, M. 1969. The Experimental Determination of Indifference Curves (*The Review of Economic Studies*, 36, 433–451) is a very explicit response to various criticisms leveled against precursor experiments in this area. Indeed, these pages demonstrate convincingly the seriousness with which methodological issues were being discussed at the tail end of the 1960s. This statement is also true for Jim Friedman's paper which was published in *The Review of Economic Studies*, as part of *a symposium on experimental economics (Friedman, J. W. 1969. On Experimental Research in Oligopoly. Review of Economic Studies 36, 399–415).*

Interestingly, the related experimental work on transitivity by mathematician/mathematical economist (May, 1953, 1954), economist Papandreou (1953), Papandreou et al. (1955), and sociologist Rose (1957), extensively and superbly discussed in Moscati (Moscati, 2007, pp. 376–84), was also based on hypothetical choices and subject to the Wallis–Friedman critique. One can speculate whether these papers did not place in good journals for that reason (with the notable exception of Rose's), but Moscati argues that they were nonetheless (ironically, again with the notable exception of Rose's) influential in triggering a debate of the transitivity axiom. Moscati's case seems persuasive. (One could, of course, ask whether these papers might have been even more influential had they not been subject to the Wallis–Friedman critique.)

### *The Methodological Questions This Museum Piece Highlights*

This museum piece addresses the hypothetical nature of the stimuli and subjects' unwillingness to state or—because of the artificiality of the experimental situation—their inability to know their true reactions, the unrepresentativeness of subjects' response, and the fact that preferences in economic situations are unlikely to be stable.

## Episode Two: Morgenstern (1954) on Experiment and Large-Scale Computation in Economics

### *Context*

> The possibilities of controlled direct experiments in the economy as a whole are very numerous—contrary to a widespread belief of the opposite. Indeed, they are only limited by the amounts of money one wishes to devote to them and by restrictions of ethics, common decency, political prejudices and the like—all of them very sound restrictions. However, even within these restrictions a larger monetary effort could provide significant quantities of new information not available so far. (Morgenstern, 1954, p. 515)

Innocenti and Zappia (2005, pp. 84–85) point out that in his earlier book on the reliability of economic data, Morgenstern (1950) listed among the many sources of errors in economic statistics the lack of verification through experiments. In fact, it was the first of the sources that he listed. Morgenstern (1954) is a reversal of sorts because there he makes the persuasive case—to which later authors referred (e.g., Kagel & Winkler, 1972; Kagel & Battalio, 1980)—that experiments were a natural for economics. His revised assessment was quite likely a consequence of Morgenstern attending the University of Michigan Summer 1952 Seminar at Santa Monica which brought together a diverse group of researchers. This gathering strikes me as the most important event for experimental economics during the

1950s, an assessment shared by others (e.g., Heukelom, 2010). In Thrall et al. (1954, p. 331), Morgenstern is listed as having contributed a paper titled "Experiment and Computation in Economics" which seems to have been a precursor of Morgenstern (1954) which was written in 1953 (Morgenstern, 1954, p. 496).

It is noteworthy, particularly in light of Naylor's revisitation[4] of Castro and Weingarten (1970), to note that to Morgenstern computations could be substitutes for experiments (Naylor, 1972). Foreshadowing the purposes of experiments later identified by witness seminar participants (e.g., Roth, 1995), he states:

> We distinguish two types of experiments: (1) Experiments of the first kind are those where new properties of a system are to be discovered by its manipulation on the basis of a theory of the system; (2) Experiments of the second kind do not primarily rely on a theory but aim at the discovery of new, individual facts. The distinction is not sharp, since the results of the experiments of the second type are eventually incorporated into a theory whereby they receive their standing.
>
> We can now state a general thesis: Every computation is equivalent to an experiment of the first kind and vice versa. The equivalence rests on the fact that each experiment (certainly each of the first kind) can be conceived of as being—or using—an analogue computing machine (Morgenstern, 1954, pp. 499–500).

This thesis is then expanded on a dozen pages. In the following summary I shall focus on the section in Morgenstern (1954) which follows those pages and which deals with experiments in today's meaning of the word.

[4] Castro and Weingarten set out to explain "the manner in which experimental techniques may be systematically applied to the analysis of basic economic behavior" (Castro, B. & Weingarten, K. 1970. Toward Experimental Economics. *Journal of Political Economy*, 78, 598–607). Both Kagel and Naylor reacted explicitly to the Castro and Weingarten article, the former pointing out the "advantages and potentialities of designed experimentation in token economies" (Kagel, J. H. 1972. Token Economies and Experimental Economics. *Journal of Political Economy*, 80, 779–785), by way of several examples, while the latter argued "[the] paper by Castro and Weingarten is both misleading and incomplete. First, the literature in economics abounds with articles on experimental economics. Second, the authors' survey of the literature in experimental economics omits some of the most important contributions in this field. Third, the authors completely ignore an important new field of experimental economics – namely, computer simulation experiments with models of economic systems. My paper contains an extensive survey of the literature on real-world experiments in economics as well as computer simulation experiments" (Naylor, T. H. Ibid. Experimental Economics Revisited. 347–352). Among the articles Naylor discussed were a complete list of those motivated by Thurstone (1931) including MacCrimmon and Toda (1969), multiple references to Siegel's work, more than a dozen articles on market experiments (missing, though, the work of the "Frankfurt school"; see Sadrieh, A., Abbink, K. & Tietz, R. 2008. *Experimental economics in Germany, Austria, and Switzerland: a collection of papers in honor of Reinhard Tietz*, Marburg, Metropolis-Verl.), half a dozen articles dealing with large-scale social experiments, and a long section on computer simulation experiments.

## *Summary*

Morgenstern acknowledges "first the occasional appearance of strictly planned experiments and second the ability to compute on a large scale (with the aid of electronic computers) by making use of currently available theory. … During the current decade still further possibilities will undoubtedly be explored of which those connected with experiment and computation appear to be especially promising. Their particular appeal lies – at least to my mind – in the combination of a profound study of the data and their new processing, with a rigor of the theoretical reasoning that can compare favorably with that of the natural sciences" (Morgenstern, 1954, pp. 484–5). Clearly the natural sciences (explicitly physics and astronomy) were Morgenstern's template. He called them "the advanced empirical sciences" (Morgenstern, 1954, p. 485) and left little doubt that economics, to his mind, was an empirical backwater.

Acknowledging that making experiment and large-scale computation standard tools in the profession's toolbox would not be easy because economists had to acquire new skills and to become acquainted with new ideas and techniques, Morgenstern left no doubt that the conventional wisdom regarding the impossibility of experimentation in economics was wrong: "I do believe that there exists great opportunity for direct experiments now and in the future. I am thinking of the actual, physical, experiment, i.e., one in which physical reality is being subjected to desired conditions, as distinguished from the so-called 'thought experiment'" (Morgenstern, 1954, p. 486); he labels thought experiments "indirect" experiments and the actual physical experiments "direct" experiments. His concern is mainly with the latter and our ability "in the real, physical, world under controlled conditions [to change] those variables that economists deem significant in their science and upon which one may be able to operate" (Morgenstern, 1954, p. 487).

After a lengthy section in which Morgenstern summarizes the development toward computation and computability in economics (touching on issues such as the solvability and stability of systems of many equations, the lack of experimental and other empirical determinations of initial conditions, and parameters, knowledge requirements, and computation techniques), he turns his attention to "the direct experiment and measurement" (Morgenstern, 1954, pp. 506–511) and "experimental possibilities in economics" (Morgenstern, 1954, pp. 511–520: direct experiments, pp. 520–538), stressing that "the range of direct experiments thus far performed in economics or feasible in the future is very considerable. The frequently encountered opinion that direct experiments (…) for all practical purposes are impossible, cannot be maintained. On the contrary, the possibilities are numerous and depend to a large extent merely on the (monetary) means to be utilized" (Morgenstern, 1954, p. 512).

Arguing that "a history of the chief economic experiments performed would be of high value and very instructive" (Morgenstern, 1954, p. 512), he then summarizes those he knows. He talks about von Thuenen's agricultural experiments,[5] arguing that "Thuenen's combination of experimental and theoretical effort has never been matched or surpassed in economics" (Morgenstern, 1954, p. 513), and the routine marketing experiments done by individual business organizations (Morgenstern, 1954, p. 513),[6] arguing that they represent a lost opportunity for significant advances, and he recommends "a fine experiment" (Brunk & Federer, 1953) and some literature quoted in that article as templates for good experimentation.

He then argues that "the possibilities for experimentation in business are practically inexhaustible: management can experiment with wage rates, hours, pay systems, etc. in normal surroundings far in excess of the ideas management has now, and economics could profit from the results immediately provided the cooperation is established which is indispensible for the progress of both" (Morgenstern, 1954, p. 514).

In a similar vein Morgenstern then stresses the ample opportunities for experimentation ("policy measures") that government has. It does not use them in satisfying ways, though; "as experiments they are unclean, vaguely conceived and inadequately described so that the possibilities of an exploitation of the experiences for scientific purposes are very limited" (Morgenstern, 1954, p. 515).

Morgenstern finally dives into his account of designed experiments involving large aggregates, mentioning the sizable Jesuit settlements in Paraguay in the seventeenth and eighteenth centuries, Fourier's "Phalanxes" in France, Owen's organizations in Scotland, various religious communities in the United States that employed special economic systems, the dated money experiment of the Woergl community in Austria, and the social credit movement in Canada. He sees these attempts as valid templates of designed experiments, essentially real-life laboratories that could speak to various issues (price systems, monetary arrangements, etc.). He even addresses the issue of the remuneration of the participants in these experiments: "Their remuneration introduces complications as well as simplifications. Both have been encountered previously and can be duly considered (cf. the Mosteller-Nogee experiment where essentially the same situation arose)" (Morgenstern, 1954, p. 517).

Morgenstern then goes on to discuss the possibilities of token economies of various kinds (more on this below). He singles out Radford's "most interesting and

[5] He "carried out extensive and costly experiments on his large estate, kept precise records, designed new methods for experimentation, all this lasting many years" (Morgenstern, O. 1954. Experiment and Large-Scale Computation in Economics. *In:* Morgenstern, O. (ed.) *Economic Activity Analysis*. New York: John Wiley).

[6] "Good examples are offered by a company planning the sale of new product and testing in similar, but separate towns, different advertising campaigns using various packages, prices, timing, etc. ... These experiments, so far as I can see, are totally neglected in economics where they are, or rather will be, of the greatest significance, even in their present form. ... Steps should be taken for a close cooperation with business firms; the benefits would be mutual and of the highest significance" (ibid).

brilliant account of price fluctuations (mostly in terms of cigarettes) in a British P.O.W. Camp" (Radford, 1945) as a prototype of experimental studies that "may reveal decisive properties of demand, money, preferences, etc. about which we suspect nothing to at present" (Morgenstern, 1954, p. 517). He even discusses the problems of experimenter effects arguing that, while important to deal with, they are not an obstacle to the possibility of experimentation (Morgenstern, 1954, p. 518).

Moving on to designed experiments not involving large aggregates, and stressing that his knowledge is limited to publications, Morgenstern discusses skeptically (Chamberlin, 1948) "who used students in a course on economics to construct a market, mostly for pedagogical purposes, although some new results are also claimed" (Morgenstern, 1954, p. 519). He next discusses, much more positively, the (Mosteller & Nogee, 1951) experiment. He sketches its purpose—also relative to Neumann and Morgenstern (1944)—and judges its merits as follows: "This may be the first direct experiment in economics that can compare with those in the physical sciences, including psychology. It is a true experiment and goes in every respect far beyond the questionnaire . . ." (Morgenstern, 1954, p. 519). The reference to the Mosteller–Nogee experiment and his assessment are interesting because they strongly suggest that Morgenstern must have been fully aware of the Wallis–Friedman critique of Thurstone's experiment.[7] In fact, throughout his article, one gets the impression that Morgenstern has bought into that critique hook, line, and sinker and that he takes the methodological implications of that critique for a given.

Morgenstern also discusses Ward Edwards' related experiment on economic decision making in gambling situations, pointing out that it had been presented at the 1952 Meetings of the Econometric Society. He stresses the relevance of this psychologist's work for economics because of its both relevant subject matter and methodological "neatness" (Morgenstern, 1954, p. 520). He concludes: "From these experiments—and the many that will undoubtedly follow—will result a theory of utility that of a truly scientific character, removed from the realm of pure speculation" (Morgenstern, 1954, p. 520).

He concludes his discussion of direct experiments with a reference to "quite extensive experiments" on games of strategy, referencing explicitly the Kalisch et al. (1954) Rand Corporation working paper that was published as Chap. 19 in Thrall et al. (1954) the same year (Kalisch et al., 1954). Interestingly, he argues "[these experiments] aim at gaining information about tendencies to form coalitions, their stability, preferences for certain types of strategies, etc. As long as these particular games are not specifically identified with typical economic situations we shall not enter upon further discussing, although they are potentially very important" (Morgenstern, 1954, p. 521).

[7] Mosteller and Nogee do not cite Wallis and Friedman (1942), but they mentioned prominently and repeatedly (footnotes 1 and 7) that their experimental design benefitted from discussions with Friedman and Savage, as well as Wallis (Mosteller, F. & Nogee, P. 1951. An Experimental Measurement of Utility. *The Journal of Political Economy*, 59, 371–404). Throughout their article one finds methodological reflections that are clearly responses to the objections that Wallis and Friedman (1942) enumerate.

## *The Methodological Questions This Museum Piece Highlights*

Morgenstern's chapter is remarkable in that he bluntly, confidently, and—in light of the considerable evidence that he lays out—convincingly contradicts what he identifies as the prevailing opinion at that time: that controlled direct experiments are not implementable. In contrast, Morgenstern sees them as a necessary step to move away from the shaky empirical foundations that he sees the house of economics built on (e.g., Morgenstern, 1954, p. 489). He also seems to have accepted the Wallis–Friedman critique and seems to have methodological conceptions (financial incentives, ethical restrictions—probably implying no deception—external validity (if necessary through work with token economies; see Morgenstern, 1954, p. 517)) that seem strikingly in line with today's conceptions—at least experimental economists' conceptions (e.g., Hertwig & Ortmann, 2001)—of what constitutes appropriate forms of experimentation.

# Episode Three: Thomas Juster (1970) on the Possibilities of Experimentation and the Quality of Data Input in the Social Sciences

## *Context*

Juster's article mirrors Morgenstern's concerns about the quality of data accessible to, and produced by, social scientists, how and why the production of economic knowledge differs between natural sciences and social sciences and what, if anything, can be done about it; it seems not coincidental that Juster refers to the second edition of Morgenstern (1950) on the first page of this article (Juster, 1970, footnote 3, see also footnote 11). They were clearly in agreement that "economists possess a very large and often quite useful stock of qualitative knowledge but a remarkably skimpy stock of quantitative knowledge" (Juster, 1970, p. 139). The specific concern of both was the reliance on the analysis of existing data, at sharply diminishing returns, and the need for new "experimental sets of microdata" (Juster, 1970, p. 138). Morgenstern and Juster had different interpretations of the word "experimental," though. As we will see, for Juster "experimental" had to do with the framing and treatment definitions of survey questions. Incentive-compatible elicitation does not come up as an issue.

Juster addressed his concern in a series of articles before (e.g., Juster, 1960, 1961, 1964, 1966, see also Brady, 1965; Namias, 1965) and after 1970 (e.g., Juster, 1974; see also Juster & Stafford, 1991, for an interesting turn to a new but clearly related area).

Some of Juster's work is related to that of George Katona of the Survey Research Center (SRC) at the University of Michigan, the same place where mathematical psychologists Coombs and Edwards resided although the interaction of Katona with

Coombs and Edwards seems to have been minimal, or nonexistent (Heukelom, 2010). The SRC was set up after the Division of Program Surveys of the Department of Agriculture was dissolved and its senior officials (including George Katona) reconstituted themselves as the SRC. The SRC transformed a survey on liquid asset into annual surveys on consumer finances that were administered by the SRC and sponsored by the Federal Reserve Board starting in 1947.

It was the Federal Reserve Board that commissioned the Smithies Report, a detailed review of the work of the SRC (1955). The report addressed many of the issues that Juster subsequently addressed "experimentally," and for a public document, it was fairly critical of the operation that Katona and his colleagues ran, as was Juster (e.g., Juster, 1961).

## Summary

The work of Juster (1964) is a study of the interaction of anticipated and actual purchases of 13 consumer durable products that ranged from new automobiles to garbage disposal units. Remarkably, this study draws on interviews, and reinterviews, of about 20,000 households from the Consumers Union membership. To understand the ambition of this project, it is useful to know that the SRC relied initially on 3500 samples and later on 3000.

*Consumer Buying Intentions and Purchase Probability: An Experiment in Survey Design* (Juster, 1966) was a postscript of sorts to *Anticipations and Purchases: an analysis of consumer behavior* (Juster, 1964).

Juster's "experiments" in survey design were motivated by the insight that consumer purchase intentions were insufficient predictors of purchase rates. The key problem was that purchase intentions are binary and that, while they work reasonably well for intenders (duly controlling for response biases, income, assets, time frame, kind of product, etc.), they tend to fail for nonintenders which for the products under consideration were naturally the vast majority. Juster's results for intenders suggested that buying intentions were reflecting respondents' subjective probability of purchasing a product and that a stated intention was a function of that subjective probability. Intentions surveys could not detect movements in mean probability among nonintenders, and, given their (nonintenders') weight in the overall sample, intentions surveys were therefore bound to run into trouble. Juster suggested that probability statements (which in his view were in any case underlying expressed purchase intentions) "might well be obtainable empirically" (Juster, 1966, p. 658) and correspondingly proposed a reframing of the questions asked in consumer surveys.

Contemporary reviewers of Juster (1964) such as Namias (1965) and Brady (1965) acknowledged his contributions to the art of analysis of survey data, research methodology, and the improvement of relevant survey techniques: "Among his contributions is the analysis of the significance of changes in wording, an essential requirement for the proper interpretation of survey data, and for increased accuracy of prediction" (Namias, 1965, p. 109). And Brady (1965, p. 203) adds, "The study is more generally a demonstration of a methodology for the study of formulation of survey questions. Through the discovery of a hypothesis about the nature of the responses to different questions on the same subject, the directions of further work can be guided efficiently with an expectation of progressive improvement in technique."

The quite dramatic improvements in the quality of predictions resulting from well-designed surveys demonstrated in Juster (1964, 1964, 1966) clearly informed (Juster, 1970, see also Juster, 1984, for a similar, even starker example). The latter was an attempt at policy intervention squarely aimed at those in power. As mentioned, the issue was the remarkably skimpy stock of quantitative knowledge and the need for new "experimental sets of microdata" (Juster, 1970, p. 138). It should be clear though that Juster was not talking about the kind of direct experiments that Morgenstern had talked about, notwithstanding the curious excursion on pp. 142–144 which seems heavily influenced by Morgenstern's ruminations on direct experiments. Compare, for example, Juster's discussion of "research designs widely used in analysis of managerial decision making and marketing strategy" (Juster, 1970, p. 142).

Juster also addresses the issue how economists got stuck in the rut of a remarkably skimpy stock of quantitative knowledge[8] and how they might manage to get out. He talks about how research economists spend their time, the costs of basic and processed data, and the high costs of new data and data input in the social sciences: "For the most part, the data inputs into economic research consist of processed rather than basic data, and economics is probably unique among the sciences in the proportion of professional resources that go into the processing and manipulation of basic data" (Juster, 1970, p. 140). He then compares this situation with the situation of the physical sciences: "Empirical research in the physical sciences is based almost entirely on observations generated as an essential part of the research process itself, and a large proportion (probably more than half) of professional

[8] "The production of new microdata sets is extremely costly. . . . The enormous cost of generating new microdata sets thus means that economists tend to use the best existing sets of data even though they may be seriously deficient for research needs, a tendency that can be offset only if the high costs are balanced by equally high returns" (Juster, F. T. 1970. Microdata, Economic Research, and the Production of Economic Knowledge. *The American Economic Review*, 60, 138–148).

"Assume that you had a dissertation student writing on the question: 'Other things equal, are retail prices higher in ghetto areas than suburban areas?' Further assume that data already available could be used to examine the problem, but that the data were, as usual, seriously deficient . . . Further assume that an extra two years would be required to collect and analyze the relevant data, and that additional costs would be incurred to raise the required funds. Given these facts, do you recommend that your dissertation student obtain the relevant observations or use existing data? I suggest that your student's professional reputation and expected income would be unaffected by the choice of research strategy, and hence that you could not in fairness recommend the investment of an extra couple of years in generating observations. Yet the social value of the two research designs is markedly different: one has the potential for providing a solid basis for policy decisions; the other does not" (ibid).

skills is devoted to the questions of what observable phenomena are to be measured and how can the measurement be made" (Juster, 1970, p. 141).

### *The Methodological Questions This Museum Piece Highlights*

The specific concern of both Morgenstern and Juster was the problematic reliance on the analysis of existing data and the need for new experimental sets of microdata. The one obvious methodological insight of relevance to experimental economics one can take home from Juster (1964, 1966) is the impact that wording can have.

## Episode Four: Token Economy and Animal Models for the Experimental Analysis of Economic Behavior (Kagel & Battalio, 1980)

### *Context*

As also evidenced in the frequent references to the work of Skinner, Ayllon and Azrin, Kazdin, and Bootzin in their earlier work during the 1970s, as well as Battalio and Kagel and their various frequent collaborators, were very familiar with the history of behavior modification.[9] Kazdin (1978) has 60 pages of references and 1200 names indexed. When Kagel and Winkler (1972) laid out areas of cooperative research between economics and applied behavioral analysis—a prospective collaboration that, for the record, they called behavioral economics—token economies and animal experiments were widely used in the behavioral sciences but not (yet) in economics.

[9] Rutherford and Lemov have provided us with highly readable accounts of the history of the many players and experiments in this space. Rutherford attempts to sketch out the huge influence that behaviorism had and, notwithstanding the naysayers, allegedly still has both within the academy and outside of it. In many respects, Rutherford's book complements Lemov's masterful study of the motivations (attempts at engineering human and social behavior on an ambitious scale) and drivers (general funding through various covert (e.g., the Human Ecology and related satellite projects) and not-so-covert state agencies (e.g., the CIA)). In Lemov's book Skinner takes second place to influential (at their time) researchers such as Jacques Loeb, John B. Watson, Beardsley Ruml, Clark Hull, O. Hobart Mowrer, Neal Miller, John Dollard, George Peter Murdock, Louis Jolyon ("Jolly") West, Harold George Wolff, and Ewen Cameron. Both authors overlook the influence that their protagonists had on experimental economics, or at least that part of experimental economics that was attracted by token economies and animal experiments. Lemov, R. M. 2005. *World as laboratory: experiments with mice, mazes, and men,* New York, Hill and Wang, Rutherford, A. 2009. Beyond the box B.F. Skinner's technology of behavior from laboratory to life, 1950s–1970s.

Interestingly, Kagel and Winkler (1972) motivated their proposal for a behavioral economics in ways similar to Morgenstern. In fact, the second edition of Morgenstern (1950) and Juster (1970) is explicitly referred to. As Kagel and Winkler (1972, p. 337) state succinctly, "there is a fundamental imbalance in behavioral economics between work on a slowly growing but still weak observational foundation and a proliferating super-structure of observationally uninterpreted theories and tedious arithmetic computational techniques. One need not look far for the reinforcement contingencies sustaining this research behavior." Empirical analysis, it is then argued, is less regarded than formal mathematical reasoning, and creating your own data is more costly than using existing data even if the latter are afflicted with various inaccuracies. Acknowledging that other options exist (and in fact pointing at Morgenstern's discussion of direct experiments and Juster's suggestions for experimental control in economic research), Kagel and Winkler make the argument for both token economies and laboratory studies of "the behavior of animals below the human level" (Kagel & Winkler, 1972, p. 339).

The work of Kagel and Battalio (1980) is a snapshot of their accomplishments in the decade afterward (e.g. Battalio et al., 1973, 1974, 1981a, b, to name a few); this chapter has been chosen as exhibit because the authors discuss their work on token economies and laboratory studies with animals below the human level in parallel and because they place emphasis on methodological issues.

McDonough, pointing out that animal experiments are barely mentioned in Kagel and Roth (1995), has pointedly asked how come that "despite confident conclusions and an admirable publication record in leading economics journals, Kagel and his associates have attracted few, if any colleagues into animal labs? . . . This failure appears mysterious" (McDonough, 2003, p. 401). Indeed, Battalio et al. (1991) was the last major publication coming from the Texas A&M rat lab, notwithstanding Kagel's persuasive plea (Kagel, 1987, see Loomes, 1988).

As a topic of interest to economists, token economies went into a tailspin already by the end of the 1970s, notwithstanding Kazdin's claim that the future was still bright. Rutherford (2009, p. 65, pp. 74–77, and Chap. 4) (see also Kazdin (1978, pp. 346–72) and Kazdin (1982, pp. 437–41)), attributes this development to a number of practical, ethical, political, and legal issues. Surely, cost must have played a role, too.

## *Summary*

Kagel and Battalio (1980) lead into their article with a figure that presents data on the labor supply behavior of a rat working for alcohol in an operant conditioning chamber and humans working for alcohol in a simplified token economy. They stress that this was apparently "the first exemplification of such a relationship using data for individual workers to appear in literature" (Kagel & Battalio, 1980, p. 380).

After discussing how to bring behavior into the laboratory (and hence addressing the issue of external validity) and why, and based on what assumptions, the small worlds of the human and animal lab can be used to test properties of the Slutsky–Hicks labor supply model, they make a methodological detour addressing the issue of why the economic behavior of nonhumans is indeed worth studying.

They argue:

> we proceed under the assumption that there is behavioral, as well as physiological, continuity across species, and if we identify genuine instances where this continuity breaks down, we shall have obtained a great deal of insight into the functional, or evolutionary, basis of the behavior in question.... This continuity in behavioral processes across species enables us to exploit the fact that the economic cost of experiments with nonhumans is considerably lower than the cost of comparable research with humans. ... Further, experiments using animals permit a degree of control and manipulation of experimental conditions that may be necessary for investigating some hypotheses but which are unethical or illegal when applied to humans. For example, tests of the hypothesis put forward by Stigler & Becker (1977) that differences in 'tastes' between individuals at a point in time are a function of differences in behavioral histories (analyzable within a neoclassical framework), rather than differences in genetic make-up, require enforced separation of parents from offspring. Such studies can only be performed using laboratory animals. (Kagel & Battalio, 1980, p. 384)

Having discussed further the use of individual observations in hypothesis testing and theory development (and in passing—footnote 8—having made clear that token economies and laboratory animal studies almost automatically address the Wallis–Friedman critique), Kagel and Battalio address internal and external validity as criteria for experimental results, the former one addressing the issue whether indeed treatments have made a difference and the latter addressing the issue of generalizability: "To what populations, settings, and variables can the effects reported be generalized?" (Kagel & Battalio, 1980, p. 390).

With the subjects typically employed in token economies and laboratory animal studies being "unrepresentative," the issue of the generalizability to "typical" behavior in national economies becomes important. Kagel and Battalio argue that trade-offs have to be made and point at their studies of consumer demand behavior in a token economy with long-term psychiatric patients (Battalio et al., 1973, 1974) which could not have been done with other human communities due to the available budget. In the following section, they give additional examples, arguing that one of them was "the closest approximation to date of Oskar Morgenstern's (1954) notion of establishing communities for the explicit purpose of conducting economic experiments" (Kagel & Battalio, 1980, p. 394). They also point to the disciplining and confidence-building role of systematic and direct replication, preferably through other researchers.

In a couple of pages later, they implicitly address the Wallis–Friedman critique of Thurstone's experiment, arguing that token economies and laboratory animal experiments address all concerns: "Experiments in preference theory using individual subject data as the unit of observation are not new to economics (May, 1954; Rousseas & Hart, 1951; Thurstone, 1931). What is new about our experiments is that the technologies employed result in the commodities and/or jobs in the choice set being an integral part of the ongoing activities of subjects for reasonably long periods of time. This automatically induces nontrivial values on the outcomes of individual responses to the experimental contingencies, an important element in effectively designing economics experiments (Siegel, 1961; Smith, 1976) and a serious deficiency in earlier experimental studies (see MacCrimmon & Toda, 1969, for citations to this earlier literature)" (Kagel & Battalio, 1980, p. 386).

As regards internal validity, Kagel and Battalio discuss the importance of within-subject replication (ABA designs) and warn that one ought not to take for granted reversibility to the baseline A condition. The occasional failure of that reversibility makes necessary refinements in experimental methods and theory.

### *The Methodological Questions This Museum Piece Highlights*

Token economies and laboratory animal settings, at least to the extent that they can be cleanly designed and implemented, are microcosms that address the methodological criticism that Wallis and Friedman directed at the Thurstone experiment. In fact, the arguments in favor of these two then new (for economics but by no means for biology and psychology) technologies in the experimental analysis of economic behavior seemed very persuasive indeed.

## Episode Five: Siegel's Work on Guessing Sequences

### *Context*

> I have not the slightest doubt that if Sid Siegel had lived, say another 25–30 years, the development of experimental economics would have been much advanced in time. He was just getting started, was a fountain of ideas, a powerhouse of energy, and had unsurpassed technique and mastery of experimental science. Twenty-five years later I asked Amos Tversky, 'Whatever happened to the tradition of Sidney Siegel in psychology?' His answer: "YOU'RE IT!". (Smith, as quoted in Hertwig & Ortmann 2001, p. 442)
>
> Sid was more than a master experimentalist, he also used theory and statistics with skill in the design and analysis of experiments. I am persuaded that if Sid had lived he would not only have been the deserving Nobel Laureate who was well out in front of the rest of us, but also the timetable for the recognition of experimental economics would have been moved up perhaps several years. (Smith, 2008, p. 198)

Simon (1959, e.g., pp. 258–9) discussed empirical studies on decision making under uncertainty, noting that the new axiomatic foundations of utility theory "have led to a rash of choice experiments. An experimenter who wants to measure utilities, not merely in principle but in fact, faces innumerable difficulties. Because of these difficulties, most experiments have been limited to confronting the subjects with alternative lottery tickets, at various odds, for small amounts of money. The weight of evidence is that, under these conditions, most persons choose in a way that is reasonably consistent with the axioms of the theory—they behave as though they were maximizing the expected value of utility and as though the utilities of several alternatives can be measured. [Here footnote that refers to Edwards, 1954[10] and Davidson et al., 1957]" (Simon, 1959, p. 258).

This assessment, and also his assessment of the literature on transitivity in the following footnote, presented a problem of sorts for Simon (1959) and his campaign against classic theory (e.g., Simon, 1947; Simon, 1957). His way out was squarely aimed at the external validity of the available evidence:

> When these experiments are extended to more 'realistic choices—choices that are more obviously relevant to real-life situations—difficulties multiply. In a few extensions that have been made, it is not at all clear that the subjects behave in accordance with the utility axioms. There is some indication that when the situation is very simple and transparent, so that the subject can easily see and remember when he is being consistent, he behaves like a utility maximizer. But as the choices become a little more complicated—choices, for example, among phonograph records instead of sums of money—he becomes much less consistent. [References to Davidson et al. (1957) and May (1954), with the footnote referring to Rose (1957) and the published version of Papandreou (1953); see (Moscati, 2007, p. 379)]

The external validity critique is also implicit when later in his paper, Simon addresses binary choice experiments, arguing that much recent discussion about utility had centered around a particularly simple choice experiment: "This experiment, in numerous variants, has been used by both economists and psychologists to test the most diverse kinds of hypotheses. ... How would a utility maximizing subject behave in the binary choice experiment? Suppose that the experimenter

[10] Interestingly, psychologist Ward Edwards was the discussant of the May and Papandreou papers at the Chicago 1952 meeting of the Econometric Society. Edwards had just completed his Ph.D. at Harvard under John G. Beebe Center and Mosteller, his dissertation being "an experimental study on risky choices that built on the Mosteller-Nogee experiment" (Moscati, I. 2007. Early Experiments in Consumer Demand Theory: 1930–1970. *History of political economy.*, 39, 359–402). Let us add that Ward Edwards, by his own recognizance (Edwards, W. 1981. This week's citation classic (W Edwards, 1954). *Citation Classics,* 6, 312.), as a graduate student read an article by psychologist Kurt Lewin and coauthors about aspiration levels and in the same week attended a seminar where Mosteller talked about his work:

"The obvious similarity of the ideas from such dissimilar roots stimulated me, and interacted with knowledge of economic theory obtained because my father was a prominent economist and because I had taken many economics courses as undergraduate. This confluence of ideas (which in fact, as I learned much later, converged earlier in Berlin in 1928) led to my PhD thesis and to my lifelong research interest in decision making. At Johns Hopkins, in my first post-PhD job, I decided that I could not afford my existing state of ignorance about the literature of decision making, then mostly a topic in economics and statistics. Since I had to write the stuff anyhow, I chose to get a publication out if it by writing a review. So I borrowed for 3 months an office in the library of the economics department, holed up all day, every day, and emerged with 'The theory of decision making' in virtually final-draft form. ... The topic of decision making apparently was not salient in psychological thinking as available for research and theorizing; this article made it so. ... Economists, too, found it useful; they had not considered the possibility that the assertions about individual behavior and rationality that make up the content of microeconomics were readily subject to experimental test, and that some such tests had been performed, with dubious results" (ibid). Edwards, W. 1954. The theory of decision making (*Psychological Bulletin* 51, 380–417) became indeed a citation classic.

rewarded 'plus' one one-third of the trials, determined at random, and 'minus' on the remaining two-thirds. Then a subject, provided that he believed the sequence was random and observed that minus was rewarded twice as often as plus, should always, rationally, choose minus. He would find the correct answer two-thirds of the time, and more often than with any other strategy. Unfortunately for the classical theory of utility in its simplest form, few subjects behave in this way. The most commonly observed behavior is what is called event matching. [Reference to an example of data consistent with event-matching on p. 283] ... All sorts of explanations have been offered for the event-matching behavior. ... The important conclusion at this point is that even in an extremely simple situation, subjects do not behave in the way predicted by a straightforward application of utility theory" (Simon, 1959, pp. 260–1).

Or did they? "Decision-making behavior in a two-choice uncertain outcome situation" (Siegel & Goldstein, 1959) was published the same year. It was a remarkable article because it demonstrated that rejection or acceptance of the classical theory of utility was importantly conditioned on the way the experiment was conducted.

## Summary

In a widely cited and well-known article, e.g. Kagel and Battalio (1980), that summarized his explorations on binary choice games and the alleged evidence against the classical theory of utility, Siegel stated:

> This is a curious result. Since the subject is instructed to do his best to predict correctly which of the two events will occur, and since we may suppose that he is attempting to follow those instructions, should we not expect him to learn to maximize the expected frequency of correct predictions? To do this, he should tend to predict the more frequent event on all trials. We might expect him to come to such stable-state behavior after an initial learning period. (Siegel, 1961, p. 767)

To better understand this result, Siegel and Goldstein (1959) ran three treatments labeled no payoff, reward, and risk, with the second treatment only rewarding good predictions and the third treatment also punishing bad predictions. The results are fairly clear, indicative, and self-explanatory (Tables 9.1 and 9.2).

Demonstrating the effects of financial incentives was not the only methodological theme reflected in Siegel's work. A remarkable collection of assessments in his honor (Messick & Brayfield, 1964), published a couple of years after his untimely death, features Siegel's key articles but also contributions by his collaborators, students, and wife, psychologist Alberta Engvall Siegel, whose "memoir" touches[11] on Siegel's nonscientific and scientific accomplishments.

[11] ... and does so touchingly, yet unsentimentally ...

**Table 9.1** Number of *SS* predicting more frequent event at various proportions during final 20 trials of first 100-trial series

| | Condition | | |
|---|---|---|---|
| Proportion of predictions of more frequent event | No payoff ($N = 12$) | Reward ($N = 12$) | Risk ($N = 12$) |
| 0.60 | 2 | | |
| 0.65 | 3 | | |
| 0.70 | 3 | 3 | |
| 0.75 | 2 | 4 | |
| 0.80 | 2 | 3 | |
| 0.85 | | 1 | 3 |
| 0.90 | | 1 | 1 |
| 0.95 | | | 5 |
| 1.00 | | | 3 |

**Table 9.2** Mean proportion of times the more frequent event was predicted by a subgroup of four *SS*, randomly selected from each payoff group, during final 20 trials of each 100-trial series

| Series | No payoff | Reward | Risk |
|---|---|---|---|
| 1–100 | 0.69 | 0.78 | 0.95 |
| 101–200 | 0.74 | 0.85 | 0.95 |
| 201–300 | 0.75 | 0.86 | 0.95 |

Siegel and Andrews (1962) provided similar evidence for children

Engvall Siegel identifies the four major areas of research that Siegel contributed to as "statistics and research decisions," "measurement and decision making," "level of aspiration and bargaining," and "choice behavior."

Engvall Siegel also comments on the tenets that guided Siegel's scientific work (Engvall Siegel, 1964, pp. 17–22); she stresses that Siegel believed that research ought to be guided by theory, that exploratory studies (or what today might be called fishing expeditions) had no place in research, that data ought to be analyzed only to the extent that a prior hypothesis warranted it, that remaining close to the data was important, and that experimental work was in many situations the way to go because of the control it afforded.

Engvall Siegel (1964, pp. 19–20) first addresses the artificiality critique (Schram, 2005):

> Only by involving ourselves in our subjects' everyday lives in their own milieus can we succeed in studying the important variables, those which really make a difference. The effects an experimenter can produce through his manipulation of some independent variable in an experiment are trivial and insignificant compared to the massive effects produced by the profound manipulations which nature provided. Finally, if social scientists are to succeed in finding answers to the important social questions, they will need to observe the phenomena of interest in their genuine context. Sid's work exemplifies replies to these arguments. In the first place, the term 'experimentation' refers to a design and not a location. The essential features of experimental design are control and comparison, with randomization an essential part of control. Where randomization can be achieved in the

> field, experiments can be conducted away from any laboratory, ... (Engvall Siegel, 1964, pp. 19–20)
>
> Second, experiences in a laboratory need not be artificial and removed from putative 'real life'. With ingenuity, laboratory experiences may be arranged which are meaningful for the subjects and in which they become personally absorbed. At present, the technique for making laboratory experiences meaningful which has vogue among social psychologists is the technique of deception. ... Sid disliked and avoided deception, principally on ethical grounds, but also because an experiment involving deception creates a climate of suspicion and distrust toward psychological experimentation. In some laboratories of social psychology, the subject enters wondering what lie he is going to be told this time. ... In Sid's, the subject entered wondering how much money he would make. I say this facetiously to introduce an alternative approach for making laboratory experiences meaningful, the one Sid employed. He built into the experimental situation features which enlisted the subject's motivation. He believed in 'the payoff'. ... [follows detailed discussion of papers in the Messick & Brayfield volume]. ... (Engvall Siegel, 1964, p. 20)
>
> The important feature of the payoff is ... that the amount of the payoff to the subject depends directly and differentially on how the subject performs in the experiment; ... Sid's convictions about decision making convinced him that meaningful observations of social behavior could be made best where payoffs are involved, and he searched for situations in which a payoff could readily be employed. He was suspicious of any study whose measure depended on the subject's good will or cooperativeness toward the experimenter. As any psychologist would, he always laughed at the 'gedanken experiments' of economists in which they try to settle empirical questions by imagining what they would do in the economic situation under control. But he parted company with many psychologists in thinking that self-report measures—adjective checklists, self-rating scales, preferences inventories, attitude questionnaires, personality inventories, and the like—are similarly suspect. (Engvall Siegel, 1964, pp. 20–1)

## *The Methodological Questions This Museum Piece Highlights*

Siegel's work was indeed the precursor of the experimental practices that experimental economists adopted quickly (later codified in Smith, 1976; Smith, 1982). His methodological stance was affected by the Wallis–Friedman critique (which he did not cite in his major articles but which he surely knew since he referenced (Mosteller & Nogee, 1951) in Siegel (1957)). Essentially, in his own experiments, Siegel (1961) addressed all elements of the criticism that Wallis and Friedman had formulated and additional ones (e.g., concerning the noise that deception might bring into the laboratory).

## Concluding Remarks

Reading the articles/papers reviewed above was, for the most part, a thoroughly enjoyable journey. Many of them have stood the test of time remarkably well and they remain, even though they may be museum pieces, excellent reads. Throughout we see a concern with the hypothetical and otherwise artificial nature of the stimuli and subjects' unwillingness to state, or—maybe—their inability to know, their true reactions, the unrepresentativeness of subjects' response, the importance of the wording of instructions and surveys, and the fact that preferences in economic situations are unlikely to be stable. Acknowledging these issues, Morgenstern's chapter is remarkable for its blunt, confident, and ultimately correct assessment of the potential of the experimental enterprise in economics. Decades later the experimental method would become an indispensable tool in the economist's toolbox, a development that was authenticated through the Nobel Memorial Prize in Economic Sciences for Reinhard Selten (whose experimental work was explicitly mentioned in the citation), Vernon L. Smith, and Alvin E. Roth in 1994, 2002, and 2012, respectively.

## References

1955. Consumer survey statistics; report [submitted to] the Subcommittee on Economic Statistics of the Joint Committee on the Economic Report. Washington: Board of Governors of the Federal Reserve System. (Smythies Report)

Battalio, R. C., Fisher, E. B., Kagel, J. H., Basmann, R. L., Winkler, R. C. & Krasner, L. 1974. An Experimental Investigation of Consumer Behavior in a Controlled Environment. Journal of Consumer Research, 1.

Battalio, R. C., Green, L. & Kagel, J. H. 1981a. Income-Leisure Tradeoffs of Animal Workers. The American Economic Review, 71, 621–632.

Battalio, R. C., Kagel, J. H. & Kogut, C. A. 1991. Experimental Confirmation of the Existence of a Giffen Good. The American Economic Review, 81, 961–970.

Battalio, R. C., Kagel, J. H., Rachlin, H. & Green, L. 1981b. Commodity-Choice Behavior with Pigeons as Subjects. Journal of Political Economy, 89, 67–91.

Battalio, R. C., Kagel, J. H., Winkler, R. C., Fisher, E. B., Basmann, R. L. & Krasner, L. 1973. A Test of Consumer Demand Theory Using Observations of Individual Consumer Purchases. Economic Inquiry, 11, 411–428.

Brady, B. S. 1965. Book review of Juster (1964). Journal of Marketing Research, 2, 202–4.

Brunk, M. E. & Federer, W. T. 1953. Experimental Designs and Probability Sampling in Marketing Research. Journal of the American Statistical Association, 48, 440–452.

Castro, B. & Weingarten, K. 1970. Toward Experimental Economics. Journal of Political Economy, 78, 598–607.

Chamberlin, E. H. 1948. An Experimental Imperfect Market. Journal of Political Economy, 56, 95–108.

Davidson, D., Suppes, P. & Siegel, S. 1957. Decision Making: An Experimental Approach, Stanford, Stanford University Press.

Edwards, W. 1954. The theory of decision making. Psychological Bulletin 51, 380–417.

Edwards, W. 1981. This week's citation classic (W. Edwards 1954). Citation Classics, 6, 312.

Engvall Siegel, A. 1964. Sidney Siegel: a memoir. In: Messick, S. & Brayfield, A. H. (eds.) Decision and Choice. Contributions of Sidney Siegel. New York McGraw-Hill Book Co.

Friedman, J. W. 1969. On Experimental Research in Oligopoly. Review of Economic Studies 36, 399–415.

Hertwig, R. & Ortmann, A. 2001. Experimental practices in economics: a methodological challenge for psychologists? The Behavioral and Brain Sciences, 24, 383–403.

Heukelom, F. 2010. Measurement and decision making at the University of Michigan in the 1950s and 1960s. Journal of the History of the Behavioral Sciences, 46, 189–207.

Innocenti, A. & Zappia, C. 2005. Thought- and performed experiments in Hayek and Morgenstern. In: Fontaine, P. & Leonard, R. (eds.) The Experiment in the History of Economics. London: Routledge.
Juster, F. T. 1960. Prediction and Consumer Buying Intentions. The American Economic Review, 50, 604–617.
Juster, F. T. 1961. Book Review: The Powerful Consumer. Journal of Political Economy, 69, 503–504.
Juster, F. T. 1964. Anticipations and purchases: an analysis of consumer behavior, Princeton, N.J., Princeton University Press.
Juster, F. T. 1966. Consumer Buying Intentions and Purchase Probability: An Experiment in Survey Design. Journal of the American Statistical Association, 61, 658–696.
Juster, F. T. 1970. Microdata, Economic Research, and the Production of Economic Knowledge. The American Economic Review, 60, 138–148.
Juster, F. T. 1974. The Use of Surveys for Policy Research. The American Economic Review, 64, 355–364.
Juster, F. T. 1984. Basic Research in the Social and Behavioral Sciences. Science, 226, 610–610.
Juster, F. T. & Stafford, F. P. 1991. The allocation of time : empirical findings, Behavioral models, and problems of measurement. Journal of Economic Literature (Stanford), 29, 471–522.
Kagel, J. H. 1972. Token Economies and Experimental Economics. Journal of Political Economy, 80, 779–785.
Kagel, J. H. 1987. Economics according to the rats (and pigeons too): What have we learned and what can we hope to learn? In: Roth, A. E. (ed.) Laboratory experimentation in economics: six points of view. Cambridge; New York: Cambridge University Press.
Kagel, J. H. & Battalio, R. C. 1980. Token Economy and Animal Models for the Experimental Analysis of Economic Behavior. In: Kmenta, J. & Ramsey, J. B. (eds.) Evaluation of econometric models. New York: Academic Press.
Kagel, J. H. & Roth, A. E. 1995. The handbook of experimental economics, Princeton, N.J., Princeton University Press.
Kagel, J. H. & Winkler, R. C. 1972. Behavioral Economics: Areas of Cooperative Research Between Economics and Applied Behavioral Analysis. Journal of Applied Behavior Analysis, 5, 335–342.
Kalisch, G. K., Millnor, J. W., Nash, J. F. & Nering, E. D. 1954. Some Experimental n-Person Games. In: Thrall, R. M., Coombs, C. H. & Davis, R. L. (eds.) Decision Processes. New York: Wiley.
Kazdin, A. E. 1978. History of behavior modification: experimental foundations of contemporary research, Baltimore, University Park Press.
Kazdin, A. E. 1982. The Token Economy: A Decade Later. Journal of Applied Behavior Analysis, 15, 431–445.
Lemov, R. M. 2005. World as laboratory : experiments with mice, mazes, and men, New York, Hill and Wang.
Lenfant, J.-S. 2009. Experimental Indifference Curves and the Ordinalist Revolution [Online]. Available: http://hes-conference2009.com/papers/MON2D-Lenfant.pdf [Accessed April 30, 2010.
Loomes, G. 1988. Book Review: Laboratory Experimentation in Economics: Six Points of View. The Economic Journal, 98, 1206–1208.
MacCrimmon, K. R. & Toda, M. 1969. The Experimental Determination of Indifference Curves. The Review of Economic Studies, 36, 433–451.
May, K. O. 1953. The Intransitivity of Individual Preferences. Econometrica, 21, 476.
May, K. O. 1954. Intransitivity, Utility, and the Aggregation of Preference Patterns. Econometrica, 22, 1–13.
McDonough, T. 2003. Of Rats and Economists. Journal of the History of Economic Thought, 25, 397–411.

Messick, S. & Brayfield, A. H. (eds.) 1964. Decision and choice; contributions of Sidney Siegel, New York: McGraw-Hill.
Morgenstern, O. 1950. On the accuracy of economic observations, Princeton/N.J., Princeton University Press.
Morgenstern, O. 1954. Experiment and Large Scale Computation in Economics. In: Morgenstern, O. (ed.) Economic Activity Analysis. New York: John Wiley.
Moscati, I. 2007. Early Experiments in Consumer Demand Theory: 1930–1970. History of Political Economy, 39, 359–402.
Mosteller, F. & Nogee, P. 1951. An Experimental Measurement of Utility. The Journal of Political Economy, 59, 371–404.
Namias, J. 1965. What Are the Odds? Journal of Marketing, 29, 108–109.
Naylor, T. H. 1972. Experimental Economics Revisited. Journal of Political Economy, 80, 347–352.
Neumann, J. V. & Morgenstern, O. 1944. Theory of games and economic behavior, Princeton.
Ortmann, A. 2005. Field experiments in economics: some methodological caveats. Research in experimental economics: a research annual, 10, 51–70.
Papandreou, A. G. 1953. An Experimental Test of an Axiom in the Theory of Choice. Econometrica, 21, 477.
Papandreou, A. G., Hurwicz, L., Sauerlender, O. H., Franklin, W. & Brownlee, O. H. 1955. A Test of a Proposition in the Theory of Choice. Econometrica, 23, 333.
Radford, R. A. 1945. The Economic Organization of P.O.W. Camps. Economica, 12, 189–201.
Rose, A. M. 1957. A Study of Irrational Judgments. Journal of Political Economy, 65, 394–402.
Rosenthal, R. 1994. Interpersonal Expectancy Effects: A 30-Year Perspective. Current Directions in Psychological Science, 3, 176–179.
Rosenthal, R. & Rubin, D. B. 1978. Interpersonal expectancy effects: the first 345 studies. Behavioral and Brain Sciences, 1.
Roth, A. E. 1995. Introduction to Experimental Economics. In: Kagel, J. H. & Roth, A. E. (eds.) Handbook of Experimental Economics. Princeton University Press.
Rousseas, S. W. & Hart, A. G. 1951. Experimental Verification of a Composite Indifference Map. The Journal of Political Economy, 59, 288–318.
Rutherford, A. 2009. Beyond the box B.F. Skinner's technology of behavior from laboratory to life, 1950s-1970s.
Sadrieh, A., Abbink, K. & Tietz, R. 2008. Experimental economics in Germany, Austria, and Switzerland: a collection of papers in honor of Reinhard Tietz, Marburg, Metropolis-Verl.
Schram, A. 2005. Artificiality: The tension between internal and external validity in economic experiments. Journal of Economic Methodology, 12, 225–237.
Schultz, H. 1933. Frisch on the Measurement of Utility. Journal of Political Economy, 41, 95–116.
Schultz, H. 1938. The theory and measurement of demand, Chicago, Ill., The University of Chicago Press.
Siegel, S. 1957. Level of aspiration and decision making. Psychological review, 64, 253–62.
Siegel, S. 1961. Decision Making and Learning Under Varying Conditions of Reinforcement. Annals of the New York Academy of Sciences, 89, 766–783.
Siegel, S. & Andrews, J. M. 1962. Magnitude of reinforcement and choice behavior in children. Journal of Experimental Psychology, 63, 337–341.
Siegel, S. & Goldstein, D. A. 1959. Decision-making behavior in a two-choice uncertain outcome situation. Journal of Experimental Psychology, 57, 37–42.
Simon, H. A. 1947. Administrative behavior : a study of decision-making process in administrative organization, New York, The Macmillan Company.
Simon, H. A. 1957. Models of man: social and rational; mathematical essays on rational human behavior in society setting, New York, Wiley.
Simon, H. A. 1959. Theories of Decision Making in Economics and Behavioral Science. American Economic Review, 49, 253.

Smith, V. L. 1976. Experimental Economics: Induced Value Theory. The American Economic Review. Papers and Proceedings of the Eighty-eighth Annual Meeting of the American Economic Association 66, 274–279.

Smith, V. L. 1982. Microeconomic Systems as an Experimental Science. The American Economic Review, 72, 923–955.

Smith, V. L. 2008. Discovery - A Memoir, Bloomington, IN, AuthorHouse.

Thrall, R. M., Coombs, C. H. & Davis, R. L. 1954. Decision processes, New York; London, Wiley; Chapman and Hall.

Thurstone, L. L. 1931. The Indifference Function. Journal of Social Psychology, 2, 139–67.

Wallis, W. A. & Friedman, M. 1942. The Empirical Derivation of Indifference Functions. In: Lange, O., McIntyre, F. & Yntema, T. (eds.) Studies in Mathematical Economics and Econometrics in Memory of Henry Schultz. Chicago: University of Chicago Press.

Zizzo, D. J. 2010. Experimenter demand effects in economic experiments. Experimental Economics, 13, 75–98.

# Erratum to: A Witness Seminar on the Emergence of Experimental Economics

**Harro Maas and Andrej Svorenčík**

Erratum to:
Chapter 1 "A Witness Seminar on the Emergency of Experimental Economics" in: A. Svorenčík, H. Maas (eds.), The Making of Experimental Economics, DOI 10.1007/978-3-319-20952-4_1

---

Erratum DOI: 10.1007/978-3-319-20952-4_10

Due to a typesetting error, the heading of this chapter was wrong in the initial version published online and in print. The correct heading, as appears now in both the print and online version of the book, is:

A Witness Seminar on the Emergence of Experimental Economics.

We apologize for the mistake.

---

The online version of the original chapter can be found at
http://dx.doi.org/10.1007/978-3-319-20952-4_1

---

H. Maas
Centre Walras-Pareto d'Etudes Interdisciplinaires de la pensée économique et politique, IEPI, Université de Lausanne, Bâtiment Géopolis, 1015 Lausanne, Switzerland
e-mail: Harro.Maas@unil.ch

A. Svorenčík (✉)
Department of Economics, University of Mannheim, L7 3-5, 68163 Mannheim, Germany
e-mail: svorencik@uni-mannheim.de



A. Svorenčík, H. Maas (eds.), *The Making of Experimental Economics*,
DOI 10.1007/978-3-319-20952-4_10

# Endnotes

I. **Charlie Holt**: "At Carnegie all MBAs participated in a second year business game in which teams of student competed in simulated markets. These had a teaching purpose, although they had a bit of the flavor of an experiment, but with less structure and replication."

II. **Charlie Plott**: "Neither Harvey nor I had experience with experiments and we simply took the results at face value without realizing that the procedures used could have a data influencing impact. Vernon was not at Purdue, was no longer doing experiments and there was really nothing about his experiments that would have suggested the trading institutions could have such a dramatic effect. That fact was not really known until the posted price effect was discovered by Vernon and me much later."

III. An expanded version by **Charlie Plott**: "From conversations with Vernon and the experience with Harvey Reed emerged a realization that that one could take Vernon's idea about induced preferences and generalize them to induce preferences in a much broader economic environment. This broader environment involved both multiple units and possibly multiple public goods and do so with or without transfers of money. It was the world of public goods. From that realization I began to design and test theories that were evolving out of public economics, public choice, voting theory, and cooperative game theory without side payments. This environment is much different from the bargaining problem and other attempts to test propositions derived from game theory. This was a rather radical departure from markets and an even more radical departure from experimental methods of the day, which were focused on bargaining (e.g. Anatol Rapoport) and small coalitions found in social psychology and sociology."

IV. An expanded version by **Charlie Plott**: "We were studying voting equilibria, the core, bargaining sets, and doing so within institutions



A. Svorenčík, H. Maas (eds.), *The Making of Experimental Economics*,
DOI 10.1007/978-3-319-20952-4

that could be precisely defined and had never before been studied experimentally. Features of Robert's Rules led us into discovering there was a host of principles coming out of non-cooperative game theory, as opposed to classical cooperative game theory and games with side payments. The theory is based on an abstract concept of dominance (dictated by the institutions as opposed to the characteristic function of a game) and is extremely powerful in demonstrating and predicting. And I became associated with *Morris Fiorina* who had worked with *Bill Riker*, who was wondering if experiments could be used to study his theory of minimal winning coalitions in political science."

V. **Charlie Plott**: "This seems like an inappropriate comment. Bill had studied only one type of procedure and Mo had conducted no experiments prior to coming to Caltech. Neither of them was an expert on procedures." See Chap. 5, Footnote 35 for references on Riker's experimental work.

VI. **Vernon Smith**: "Charlie is referring to the joint course that he and I offered in experimental economics. Officially it was his course as I was not on the faculty. It was attended by three paying students and Jim Quirk, Lance Davis, Mo Fiorina, John Ferejohn, Bill Riker, and Roger Noll. I had my notes from the grad course I had taught at Purdue, 1963–65, and I went through those, but now we had a much bigger literature base, including political economy."

VII. An expanded version **Charlie Plott**: "That's how I got started. The papers and research with Mo Fiorina initiated established methods, procedures and results we see today in both economics and political science. Soon afterwards I invited Vernon to Caltech. He had been away from experimental research since the late 1960s but it was easy to rekindle his interest. We offered a course in experimental method and from that my attention began to get more focused on markets and market institutions. That set the stage and, of course, there are many results and discoveries that evolved from there."

VIII. **Betsy Hoffman**: This was Bernie Saffran, the Chair at Swarthmore.

IX. **Charlie Plott**: "The paper in the 1979 Russell volume "The Application of Laboratory Methods to Public Choice" is a good reference here because it shows the state of the science in the mid 1970s since the paper was written before 1978, which was the time of the conference." **Plott, Charles R.** 1979. "The Application of

Laboratory Experimental Methods to Public Choice," C. S. Russell, *Collective Decision-Making: Applications from Public Choice Theory*. Baltimore, Md.: Johns Hopkins Press for Resources for the Future, 137–60.

X. **Reinhard Selten**: "I found it very important to make Maschler's experimental work known to a wider audience. Therefore I asked Heinz Sauermann to invite him to the conference at Winzenhohl in 1977."

XI. **Charlie Plott**: "Aside from my conversations with Vernon there was no activity in experimental economics at Purdue when I was there. The lab at Purdue was a psychology lab. No one was conducting economics experiments." Compare also William Starbuck's recollection of his time at Purdue in the second half of the 1960s and his involvement in building this laboratory. **Starbuck, William H.** 1993. "Watch Where You Step!" Or Indiana Starbuck Amid the Perils of Academe (Rated Pg)," *Management Laureates: A Collection of Autobiographical Essays*. Greenwich, Connecticut; London, England: JAI Press, 63–110.

XII. **Charlie Plott**: "Cliff Lloyd was a Purdue faculty member who focused much of his research on how one might go about testing consumer theory. I think that everyone at Purdue was well aware of his ideas that consumer theory was testable and what he saw as a valid test. This is an important departure from the rest of economics because the belief of the profession at the time was that consumer theory was not testable. I think that it was the work with the rats that demonstrated how it could be tested."

XIII. **Vernon Smith**: "It was assigned routine reading at Harvard, but I took it quite seriously, and looking back it was a wonderful illustration of Adam Smith, David Hume, and Hayek's concept of an unintended—invisible hand—spontaneous order."

XIV. Expanded version **Charlie Plott**: "The interesting thing about it was that at that time, it wasn't clear what he was going to learn. We didn't realize that the procedures were overwhelmingly important. We didn't have the dominance relation—looking at it he had not connected social choice theory to a dominance relation and Carl did not think about using my voting equilibrium or the Condorcet winner as models to be tested. We really hadn't separated the idea of games without side payments from games in characteristic function form, which we did later. And so Carl did this experiment, but it seemed he didn't learn anything. But in retrospect that was a clear step towards the problem of trying to control preferences in these very complex areas where everything is of public goods nature."

XV. **John Ledyard** took Lloyd's graduate microeconomics course in his first year at Purdue and it exerted big influence on him. "He got me excited." **Vernon Smith**: "I agree; Cliff was a good example of a free spirit who bloomed at Purdue because his independence was encouraged."

XVI. Expanded and revised version **Charlie Plott**: "Now, in terms of receptiveness, let me make one comment. Our work on groups in relation to the rules they were using for decision-making was a major transition, away from the research that was taking place in the late '60s. I moved away from individuals to groups. Rather than testing theory in terms of what the individual was doing the testing as focused on system level predictions. I think that was a major transition and what we might learn from such an approach was somewhat removed from the thinking of the profession."

XVII. **Vernon Smith**: "There is a more serious issue here in that it is correct to say that experimental economics at its beginning was very closely tied with undergraduate and graduate education; it was research, yes, but also what I was teaching. Similarly at Caltech, our joint seminar in experimental economics had three undergraduates and a bunch of faculty."

XVIII. **Charlie Plott**: "They are amused because that is a realistic view of the situation then by contrast to now."

XIX. **Betsy Hoffman**: "I went to Purdue in 1981, but my third year review was in the fall of 1980."

XX. **Vernon Smith**: "The whole idea of combinatorial auctions in particular and smart computer assisted markets in general, came out of Steve's thesis and our joint papers. The first paper was **Rassenti, Stephen J.; Vernon L. Smith and R. L. Bulfin.** 1982. "A Combinatorial Auction Mechanism for Airport Time Slot Allocation." *The Bell Journal of Economics*, 13(2), 402–17.

XXI. Smith and Dolbear met through the Ford Foundation Summer Workshops in the early 1960s.
**Charlie Plott**: "I do not know Dolbear personally but I knew of his work at the time. I think that his failure to follow up the experimental work he was doing speaks to the history of how experiments were received by the profession. It might also reflect why Vernon found himself doing other things after his first experiments."

XXII. **Charlie Holt**: "My coauthored papers with Cyert and DeGroot were motivated by Cyert's experience on boards of directors and observing how decisions are made. For example, the investment paper below was based on his observation that corporate investment was driven to a large extent by the availability of retained earnings (a kind of mental accounting process) instead of be driven by interest rate measures of the opportunity cost of capital."

XXIII. **Vernon Smith:** "I once reminded Ed Prescott of this, and he said, 'Well, you experimental guys all made sure it wasn't a dead end.'"

XXIV. **Vernon Smith**: "Reading this (October 22, 2013) I checked my vita which goes back to the mid fifties and was kept up to date for annual reporting and P&T purposes. Here are my grants for experimental economics from the *National Science Foundation*: 1962–64, 1964–66, 1967–68, 1968–70, 1971–73, 1973–74, 1975–76, 1977–80, 1981–84,

1984–88, 1989–91, 1990–93, 1991–93, 1992–94, 1993–94, 1995–1997, 1995–2000, 2001–2003. I was also on various NSF panels in those years. My NSF grant, 1962–64 was based on the research that appeared in my JPE (1962) article, but the grant would have been based on a proposal written before publication because of the time lag from proposal to award."

XXV. **Charlie Plott**: "Shapley while at RAND would join us for seminars."

XXVI. **Charlie Plott**: "The question might be suggesting that things really took off with the organization of the Economic Science Association but I think that was really just a slight change in the organization and visibility that existed from Public Choice. Notice that the presidents of the Public Choice Society that followed me were all experimentalists—Vernon, Ostrom, and Ledyard."

XXVII. **Vernon Smith**: "I recall objections to the Arizona Inn—too many old people stayed there—so we moved to Westward Look."

XXVIII. **Vernon Smith**: "I was involved in the selection process; that is when I got to know Thomas Kuhn, the historian and philosopher of science, as we shared a panel together on selecting graduate students to support"

XXIX. **Vernon Smith**: "To clarify: The idea of having a specialized journal was controversial, but having a society (ESA) was not controversial. I recall, however, that there was some controversy over the name that centered on the word "Science;" some thought that it was pretentious, but others thought it expressed a simple truth about what we were up to, however pretentious it might appear."

XXX. **Vernon Smith:** "There was an earlier Caltech working paper with a title something like "Notes on some Literature in Experimental Economics" from 1973. Charlie urged me to write up something on my "theory of induced valuation" that had been around for 10 years, because he wanted something he could cite besides our long-running conversations. It was methodological on experimental economics, and not the sort of thing an economics journal was interested in. But if you got into a Proceedings AEA meeting they put you in their May issue; all that is now closely controlled so that if you are on the program it is not published except by a preapproval process."

XXXI. **Vernon Smith**: "He [McFadden] was there because he came up to talk with me afterwards. He and Gerard were quite interested in the session, and Charlie was interested in their interest. I shared NSF panel duty with Dan and he was always a supporter of experimental economics."

XXXII. **Vernon Smith**: "Charlie Holt's quote from Ed Prescott is the kind of thing you are talking about, but Ed is not going to come to a Plott seminar, stand up and say it!"

XXXIII. **Vernon Smith**: "My papers on the Groves Ledyard mechanism and the "auction mechanism" concerned a process. John Ledyard once put it well: "its not even a mechanism." But of course any implementation with real people" is going to involve a whole lot of things that are unmodelled in the "theory mechanism." I think it is fair to say that it is these issues that suckered John into experimental economics. The double auction was completely unmodelled, not even heard of in economics or in theory, when I used it to implement S&D in 1956, and after. It came from one of the old finance books **Leffler, George L.** 1951. *The Stock Market.* New York: Ronald Press. I got this book out of the Purdue Library in late 1955 and used it to inform myself on the double auction procedure which implemented my first experiment in January, 1956, and for the next fifty-odd years. Economists then were ignorant of this notion that to implement a static equilibrium theory you needed to have a procedural process for generating trades and prices. I was as ignorant as anyone except that I had been exposed (at the University of Kansas) to Alfred Marshall and Böhm-Bawerk from older traditions who were much more sophisticated, if verbal, than the equilibrium theorists. People like Herb Simon understood this in spades!"

XXXIV. **Vernon Smith**: "Bob Clower was a very independent editor who personally read the papers. Harry Johnson was like that at the JPE in the 1960s. Clower continued that tradition at the AER for 5 years and ran into really big time flak. He expected to run into it, and committed to the job for 5 years only. He pissed off all kinds of well-known economists. He would read their stuff and then not even send it out for review."

XXXV. **Vernon Smith**: "Yes, because Reinhard was very heavy into experiments by then and showed the press his lab!"

XXXVI. **Frans van Winden**: "We actually met Doug Davis in Richmond and we also met with John Kagel and Jack Ochs in Pittsburgh."

XXXVII. **Vernon Smith**: "I remember feeling that the time had come, it was no longer too early; time to get the wind blowing and it would pick up sails; we had lots of experimenters around."

XXXVIII. **Van Winden**: "I agree that it happened in Germany, but I think it was discussed in Bonn."

XXXIX. **John Kagel**: "I think the fact that Repcheck came to us gave us the impetus to put the project together as we had a backup publisher in hand and enough contributors. But we definitely did not commit to go with PUP and presented the idea to different publishers to see the best deal we could get in terms of promoting wide distribution."

XL. **Vernon Smith**: "Sid had a big influence on economists, but not on the new cognitive psychologists. At an experimental conference at Caltech in late 1980s, with both economists and psychologists in attendance, at one point I asked Danny Kahneman and Amos Tversky; "Whatever happened to the tradition of Sid Siegel in psychology?" Amos said: "You are it." He meant it as a put-down, but I felt honored! What I learned from Amos was that Sid and those of us in experimental economics were essentially Skinner behaviorists and that tradition was discredited and dead in psychology; we were some kind of anachronism from the past." See also Endnote #XLVIII.

XLI. **Charlie Holt**: "There has also been a strong focus on nonparametric analysis in the U.S., not just in Germany as Frans van Winden stated. For example, in the Methodology chapter of Davis and Holt, 1993, the *only* treatment of statistics was nonparametric, with explanations of the standard nonparametric tests like the Wilcoxon tests for various data configurations were explained with data from experiments. Similarly, the only treatment of statistics in Holt's *Markets, Games, and Strategic Behavior* (2006, Chap. 1) is also focused exclusively on nonparametric statistics. The chapters in Holt (2006) and Davis and Holt (1993) drew heavily on the path-breaking book of Sidney Siegel, *Nonparametric Statistics for the Behavioral Sciences* (1956). The Siegel book has relevant examples with data from experimental economics and psychology for all of the standard tests. The methodological influence of Siegel is indicated (you need to check on this) by the fact that the ESA best paper prize was, at least for a number of years, called the Siegel award. The influence of Siegel's work has probably been greater in experimental economics in the U.S. than on experimental psychology, although N. John Castellan—the NSF director of Decision, Risk, and Management Science Program—coauthored a revision of Siegel's original book (after Siegel's death) that updated some of the examples and terminology."

**Editor's note**: There is no evidence that the ESA best paper prize was called the Siegel Award. Holt refers to it in the ninth chapter of his 1993 book titled *Economic Behavior and Experimental Methods: Summary and Extensions*, in particular section Statistical Analysis of Data from Economics Experiments (pp. 525–528).

**References**: **Davis, Douglas D. and Charles A. Holt.** 1993. "Experimental Economics," Princeton University Press, 406–26, **Holt, Charles A.** 2006. *Markets, Games, and Strategic Behavior : Recipes for Interactive Learning*. Boston: Pearson Addison Wesley, **Siegel, Sidney.** 1956. *Nonparametric Statistics for the Behavioral Sciences*. New York: McGraw-Hill, **Siegel, Sidney and N. John Castellan.** 1988. *Nonparametric Statistics for the Behavioral Sciences*. New York: McGraw-Hill.

XLII. **Vernon Smith**: "I think you are being too charitable. I once asked someone in psychology at Purdue (Richard Swenson, later at Harvard I think) why there was so much resistance in psychology to paying subjects. He said it would divert grant funds from their other expenses including summer salary, and it was far more convenient, low cost and well established that you require all undergraduate psych majors to all be in a minimum number of experiments. So why squander funds paying them. So in the end they are homo economicus writ large."

XLIII. **Vernon Smith**: "I have long believed that we should not lie to the subjects because it is immoral to do so. But I never made that argument because I felt that the "don't spoil the subject pool" argument carried more weight. From the very beginning I found economists very receptive to the principle that deception was out of bounds. At Arizona, GMU and now Chapman we prohibit deception and it is not controversial."

XLIV. **John Kagel**: "I meant to say there are statistical techniques for determining if you have a group effect which would compromise treating your observations as independent when you reconstitute people into different groups as you would do in a voting game or in a series of auction experiments. While some of the people who claim to have only a single observation don't look at the evidence testing this hypothesis. I have a short rift on this in my Auctions chapter intended for the *Handbook of Experimental Economics*, volume 2, that Al Roth and I are putting together."

XLV. **Vernon Smith:** "These papers by List & Levitt were sometimes referred to as "Levitt or Leaveit"

XLVI. **Vernon Smith**: "I think we have not had that big of an impact on how economists think about economics. I feel a bit like Coase who has often expressed disappointment over acceptance of his approach to economics going clear back to 1937. I went last year to a conference at University of Chicago honoring Coase's 100th birthday. At Chicago they still do not understand Coase! They think Coase is about the Coase Theorem, but it is the antithesis of the Coase Theorem (which is not even in Coase)."

XLVII. **Vernon Smith**: "Sid innovated a lot more than just paying subjects. All his oligopoly treatment variation experiments were run simultaneously with hundreds of subjects, pre-randomized into the treatments. He did not want to rely on the hazards of running them in sequence and losing control of sampling variability. I think we have often found that our results are very robust and easily overcome such sampling variability, but he was right to not just assume that."

XLVIII. The conference titled *Experimental Economics and Psychology* took place in 1988 at Caltech. The attendees were: Eric Wanner (President of the Russell Sage Foundation), Colin Camerer (Pennsylvania), Donald Coursey (Washington), Robyn Dawes (Carnegie-Mellon), John Kagel (Houston), Daniel Kahneman (California Berkeley), Charles Plott (California Institute of Technology), Alvin Roth (Pittsburgh), Vernon Smith (Arizona), Shyam Sunder (Minnesota), Richard Thaler (Cornell), and Amos Tversky (Stanford). See Endnote #XL for Smith's recollection of this conference as well as Svorenčík (2016).

XLIX. **Al Roth:** "I recall that questions about motivation would be asked regularly at seminars."

L. **Vernon Smith**: "What is interesting about results when you do not pay is that they are not random and unconnected with the model, just that there is more noise and deviation, even systematic. Siegel rigorously tested payoff levels. He would add unannounced in advance an additional trial with escalated payoffs."

LI. **Charlie Plott** expanded version: "Well, I don't know about the rest of these guys, but I typically do not trust the experiments of other people and I am even skeptical of my own until I see a healthy pattern of replication and robustness. I have to do it myself. So Vernon reported that paying subjects was important (and reported the market experiment he did). I did not fully believe it was necessary so ran a series of experiments. I actually ran some with Vernon."

LII. **Charlie Plott**: "The animal research required technology. Early experimental economics could be done without it. I think that my lab was the first to receive a technology grant from NSF." See Chap. 6, Footnote 14 for the context of Plott's award in the 1990s.

LIII. **John Kagel**: "I think the story varies. NSF paid for our first animal lab equipment as it was specialized. When I went to Pittsburgh—Pitt put up the money for the computer lab plus a lab administrator; the same at Ohio State University. At Houston I think it was NSF that paid for the equipment. In all cases—NSF has paid at least something for subject payments and RAs and limited summer research support for principal investigators."

LIV. **Vernon Smith**: "I got funding for years under Jim Blackman, and no problem with money to pay subjects. It was routine before Dan came on board. But Dan followed and innovated in the tradition of Jim Blackman."

LV. **Reinhard Selten**: "In the bargaining preceding my acceptance of the offer from Bonn in 1984, I succeeded to get the right to fill these slots. At that time such rights were connected to chairs of full professors in Germany."

LVI. **Steven Rassenti**: "I first began working as a fulltime research scientist at the Economic Science Laboratory at the University of Arizona in 1984. At that time Vernon and some other faculty had been contacted to study the deregulation of wholesale electric power. That was my first project as a new employee of ESL. A project that followed soon was "Designing a competitive distribution system for natural gas pipelines" for the Federal Energy Regulatory Commission (Jan-December 1985)."

LVII. **Betsy Hoffman:** "The law firm was Sherman and Howard in Denver and the group was Brian Binger (my husband), Bob Coople, who was at Sherman and Howard, but now has his own law firm in Phoenix, and a company called Insight Research in Denver. The company was run by a married couple of Boulder psychology Ph.D.'s Renee D'Alba and Bryan Burns. I believe that Ralph D'Arge got Brian and me involved with them. Here are some papers that finally came out of that work: **Binger, Brian R.; R. F. Copple and Elizabeth Hoffman.** 1995a. "Contingent Valuation Methodology in the Natural Resource Damage Regulatory Process: Choice Theory and the Embedding Phenomenon." *Natural resources journal.*, 35(3), 443–59, **____.** 1995b. "The Use of Contingent Valuation Methodology in Natural Resource Damage Assessments: Legal Fact and Economic Fiction." *Northwestern University Law Review*, 89(3), 1029–116.

LVIII. **Charlie Holt:** "The first California greenhouse gas cap and trade auction implemented the RGGI design with some additions to curb market manipulation was conducted last week, results released Nov. 19, under the AB32 program. Bill Shobe (at UVA) and I have been running extensive experiments this fall to evaluate that program, funded by the California Air Resources Board. Here are our RGGI papers and the original consulting report (Holt et al, 2007):" **Burtraw, Dallas; Jacob Goeree; Charles A. Holt; Erica Myers; Karen Palmer and William Shobe.** 2009. "Collusion in Auctions for Emission Permits: An Experimental Analysis." *Journal of Policy Analysis and Management*, 28(4), 672–91. **____.** 2010a. "The Effects of a Loose Cap on Emissions Permits: An Experimental Analysis of Alternative Auction Formats." *Agricultural and Resource Economics Review*, 39(2), 162–75, **____.** 2010b. "Price Discovery in Auctions for Emissions Permits: A Comparison of Sealed Bid, Multi-Round, and Continuous Auctions," M. R. Isaac and D. A. Norton, *Experiments on Energy, the Environment, and Sustainability, Research in Experimental Economics, Vol. 14*. Bingley:UK, : Emerald Group Publishing, 11–36, **Goeree, Jacob K.; Charles A. Holt; Karen Palmer; William Shobe and Dallas Burtraw.** 2010. "An Experimental Study of Auctions Versus Grandfathering to Assign Pollution Permits." *Journal of the European Economic Association*, 8(2–3), 514–25, **Holt, Charles A.; William Shobe; Dallas Burtraw; Karen Palmer and Jacob Goeree.** 2007. "Auction Design for Selling $CO_2$ Emissions Allowances under the Regional Greenhouse Gas Initiative."

LIX. **Vernon Smith**: "These issues are not even remotely part of what academics and psychologists and a few economists talk about when they discuss "external validity." That conversation tends to be uninformed by any of the things being discussed in this session. If you have not been part of one of these applied exercises with practitioners you simply have no idea of what it is like to go through an initial design exercise with them, try it, redesign it with the data feedback, go again, etc. Bob Wilson once noted that the problem in the California electricity design is that it involved about a hundred people in a thinking (constructivist) exercise. In other words no real experience feedback."

LX. **Vernon Smith**: "But what we got was not "trash" just failure to replicate in a computer environment."

LXI. Expanded version **Charlie Plott**: "She should have been watching the screen because the quotes were coming across. The screen was the source of all of the information that she would be processing if she were motivated to make money. With that instruction, she immediately turned around and the market price went up.—it started to work just right. Now, you could add similar examples to that one. Another is the case in which subjects are given an incentive chart where their payoffs for units go down as units go up, and in which periods go across the page some subjects will go across read across. Basically, in such cases you are conducting an experiment that has completely different parameters than you thought. If you have an experiment where subjects are buying from the experimenter and selling to somebody else, a confused subject can buy from the experimenter and forget to sell.

LXII. **Vernon Smith**: "The main thing we learned was that it goes away with experience—all the individual error, mistakes, lack of common knowledge get corrected. Much later we found (Smith et al 2000) that asset markets converge without experience if you simply declare but not pay the dividend each period, but rather delay actual cash payment to the end. It diverts their focus away from the myopic to final future outcomes. A recent paper in AER by Kirchler et al demonstrated very nicely how instructions that focus on individual rationality can substitute for experiential learning. But some people are now questioning the implications of strong instructions that assure individual rational behavior as simply a way of guaranteeing that theory works. So it reinforces their idea that lab experiments are not useful; i.e., we already know that theory works if everyone is rational. I see it rather as examples of the kinds of error that experience can overcome, and the Kirchler stuff shows that there all sorts of teaching/training techniques that overcome those errors. "**Kirchler, Michael; Jürgen Huber and Thomas Stöckl.** 2012. "Thar She Bursts: Reducing Confusion Reduces Bubbles." *The American Economic Review*, 102(2), 865–83.

LXIII. **Charlie Plott**: "Such a design implies that if a bubble occurs, it is not because people are speculating about the preferences or decisions of others."

LXIV. **John Kagel**: "Learning takes quite some time and even then subjects have trouble generalizing what they have learned to new contexts—like increasing the number of bidders."

LXV. **John Kagel**: "Actually there are several papers dealing with this issue which are covered in my 1995 survey in the *Handbook of Experimental Economics* pp. 542–546; also see Holt and Sherman discussed on pp. 552–53."

LXVI. **Vernon Smith**: "Yes a great insight and there have been many more. In the ultimatum game, you change the context and you change the results; in some cases you get very close to sub-game perfect equilibrium. In the *Theory of Moral Sentiments*, Adam Smith tells us why: in the original versions of the ultimatum game it is a game of extortion. That is relaxed in some contexts because they define different paths to the ultimatum game situation. Big difference. See **Smith, Vernon and Bart J. Wilson.** 2015 to appear. "Fair and Impartial Spectators in Experimental Economic Behavior," *Oxford Philosophical Concepts: Sympathy* (Ed. Eric Schliesser). Oxford: Oxford university Press."

LXVII. **Vernon Smith**: "That is, typed bids to buy and asks to sell are entered anytime up to noon on Wednesday, then the bid demand is crossed with the ask supply and the trades all cleared at a single price. This process is repeated and the clearing price determined on Saturday midnight. And so on each week for 7 weeks. Traders have 3.5 days to log in survey the market history and submit bids or asks, or both, before a price and trade volume is computed and posted. See **Williams, Arlington W.** 2008. "Chapter 29 Price Bubbles in Large Financial Asset Markets," R. P. Charles and L. S. Vernon, *Handbook of Experimental Economics Results.* Elsevier, 242–46."

LXVIII. **Vernon Smith**: "It devastates the economy because people are buying long lived goods with credit, and when the prices collapse, the debt stays fixed. The down payments (margin) requirements were too loose. The resulting balance sheet crunch on households and banks stops the spending and the lending. That is what happened in 2007–8, and back in 1929–1930; the two biggest economic declines in 80 years."

LXIX. **Vernon Smith**: "We found it made a difference, if you earned the right to be the dictator. Cherry and Shogren found that Betsy, Kevin and I did not go near far enough! If you *earn the money* they are finding that if it is basically your own money, it makes a big difference. You give nothing! So dictator generosity is all about being generous with what Adam Smith called "other people's money' the OPM problem—in this case the experimenters money (EM) problem. Hence what I am saying is that the usual interpretation of the dictator game is dead in the water. As I said in my JEBO (2010) paper "What are the questions?" all kinds of standard results—maybe even some of the market results—need to be reexamined and their robustness checked. The one rule that every experimentalist should bring to the table is: No experimental results are sacred and cast in stone. Schwartz and Ang checked out OPM in 1989 in asset market bubbles. They recruited subjects under the condition that they bring their own money ($20.00) to stake themselves to see if our asset bubbles were due to OPM. They got big bubbles, but it could have gone the other way—a very important contribution that never got published. My point is that experimental economics has to come to terms with Cherry and Shogren's important findings. Take risk aversion measurement. Does it make a difference if it is your money or the experimenters? I believe neither in utility theory or probability theory as doing an acceptable job in organizing subjects decision behavior *across the spectrum* in which E(U) is applied. Maybe some of the erratic results are due to OPM."

LXX. **Vernon Smith**: "Yes, if it is the experimenters money, but if it is their own? Hardly, I think. This is the point of Cherry and Shogren."

LXXI. **John Kagel**: "The rules have now been changed—we (instructors) can request a paper instead of an exam but you need to get the whole committee to agree."

LXXII. **John Kagel**: "I also teach a joint course with a behavioral decision making psychologist from the psychology department about every other year on economics and psychology or behavioral economics. We also have a regular undergrad course in experimental economics—its an elective. We would had it for about 4–5 years now [2012]."

LXXIII. **Charlie Holt**: "I think it was 1989, before the early conference on the *Handbook of Experimental Economics*."

LXXIV. **Charlie Holt**: "the site for the web based experiments used for teaching is: http://veconlab.econ.virginia.edu/admin.php [Accessed on March 31, 2015]. Over 100,000 students have logged on to participate in these experiments in the past year (October 2011 to October 2012)"

LXXV. **Charlie Holt**: "The person who suggested this name to David Reiley was Ann Talman (my wife), he cites her in that paper."

LXXVI. **Charlie Plott**: "Almost every time I broke into a new area the research was guided by theory. Certainly that was the case for public economics and also experiments testing rational expectations."

LXXVII. **Charlie Plott**: "The essence is a clear statement of the basic principles at work."

LXXVIII. **Charlie Plott**: "I think that your point is even more general. Often the theory is so devoid of operational content it suggests nothing for you to do."

LXXIX. **Vernon Smith**: "Here are a couple of more recent papers in which you get very close to sub-game perfect equilibrium and its all about a context that reduces the extortion component in the ultimatum game: **Pecorino, Paul and Mark Van Boening.** 2010. "Fairness in an Embedded Ultimatum Game." *Journal of Law and Economics*, 53(2), 263–87, **Salmon, Timothy and Bart Wilson.** 2008. "Second Chance Offers Versus Sequential Auctions: Theory and Behavior." *Economic Theory*, 34(1), 47–67. I should add that the "extortion" interpretation of ultimatum games (second mover can reject the offer and both get zero) was suggested to Bart Wilson and me by Adam Smith's first and much neglected book. **Smith, Vernon L.** 2012. "Adam Smith on Humanomic Behavior. Edited Transcript of Keynote Address,

Academy of Behavioral Finance and Economics, UCLA." *Journal of Behavioral Finance & Economics*, 2(1), 1–20.

LXXX. This is a more detailed account by **Reinhard Selten**: "Demand inertia means the quantity sold by an oligopolist in a period is decreasing in his/her price but it also is increasing as a function of the quantity sold in the previous period. The total quantity sold in a period may depend on time expressed by the period number, but apart from this, only on the current average price. We did experiments on three-firm oligopolies with price variations and demand inertia. In every period the quantity sold by an oligopolist is a linear function of a variable called demand potential and the price of this oligopolist. The demand potential of an oligopolist is the sum of three components, first the quantity sold in the previous period, second, a multiple of the amount by which the oligopolist's price was below the average price in the previous period, and third a time component depending on the period number.

The three oligopolists had different average cost and faced a higher interest rate for negative account balances than for positive ones. The model had also some other features enhancing its complexity. It had the form of a dynamic game. My research associate Otwin Becker and I tried to find some theoretical solution, but the problem was too hard and we gave up on it. Later, I simplified the game to a theoretically tractable form and then I could determine a unique solution of the dynamic game with finitely many periods by backward induction. But then I realized that the game has many other equilibria. Nevertheless the backward induction solution seemed to me very natural and thinking about the reason for this I came up with the concept of sub-game perfectness. Thus in an oblique way an experiment on a dynamic game motivated a theoretical analysis of a highly simplified version of it and as a side effect of this analysis a basic game theoretic concept was introduced."

LXXXI. **Vernon Smith**: "Would that all theorists could be influenced by this sort of experience! This is how theory should be done; much better than making up the facts about what is important and then cranking out paper and pencil models based on them."

LXXXII. **Reinhard Selten:** "I clearly remember that these psychologists presented a paper at the 1977 conference at Winzenhohl. I looked at the papers in volume 7 of the contributions to experimental economics and come to the conclusion that this must have been the paper by Clay Hamner and Lloyd S. Baird on "The Effect of Strategy, Pressure to Reach Agreement and Relative Power on Bargaining Behavior" pp. 247–269. This paper does not even mention money left on the table. As far as I remember the authors showed an example of an experimental play in which money is left on the table by a buyer. However, no experimental play is described by a table of the paper."

LXXXIII. **Vernon Smith**: "I love you Reinhard, this is great."

LXXXIV. **Stephen Rassenti**: "Before 1981 Bell Telephone was the government regulated monopolist providing communication services. From 1976–81 I was a graduate student at Arizona who developed software and ran experiments in the Plato Lab in the Science Library."

LXXXV. **Stephen Rassenti**: "I would collect handwritten input information from all my experimental subjects, type it all into the dumb terminal, and send it into the mainframe. Once I received the results, I printed them and I showed them to the subjects by handing them the section of the printout pertaining to their individual outcomes."

LXXXVI. **Vernon Smith**: "Several papers came out of that and subsequent seminars I taught. I shocked the hell out of my colleagues by writing and publishing joint papers with undergraduates. When I wrote the first draft of Miller, Plott and Smith, I put Miller on it as co-author; he had been part of the team—I think Charlie thought I had gone beserk. Here are two papers from Arizona. The undergrads were Vannoni, Coppinger, and Titus. You cannot even imagine how thrilled Coppinger and Titus were when we won best article. Bratton was a grad but he never finished; Williams was a grad and he made history with the first electronic double auction:"
**Miller, Ross M.; Charles R. Plott and Vernon L. Smith.** 1977. "Intertemporal Competitive Equilibrium: An Empirical Study of Speculation." *The Quarterly Journal of Economics*, 91(4), 599–624, **Smith, Vernon L; Arlington W. Williams; W. Kenneth Bratton and Michael G. Vannoni.** 1982. "Competitive Market Institutions: Double Auctions Vs. Sealed Bid-Offer Auctions." *The American Economic Review*, 72(1), 58–77. **Coppinger, Vicki M.; Vernon L. Smith and Jon A. Titus.** 1980. "Incentives and Behavior in English, Dutch and Sealed-Bid Auctions." *Economic Inquiry*, 18(1), 1–22.

LXXXVII. **Charlie Plott**: "It was a psychology laboratory. No economic experiments were done there during my period at Purdue at least to my knowledge."

LXXXVIII. Expanded version **Charlie Plott**: "Vernon used quasi-linear utility functions before while we were not necessarily dealing with money payments across subjects, we were out of the world of experimental markets and into a world of only public goods. We were in a spatial world with big indifference curves drawn on pieces of paper. There are a lot of ways to do that. The charts had money attached to the indifference curves indicating the payoff to a subject should a point (different vectors of public goods) be chosen."

LXXXIX. **Charlie Plott**: "It seemed as if all of this control was not needed and in fact got in the way of what we wanted to study."

XC. **Vernon Smith**: "This reminds me that Sid Siegel used to recruit from the pool of students available for part time work around the Penn State University. He figured they were hungry."

XCI. **John Kagel**: "It would be an unofficial subtitle—spoken/not written so as not to offend people."

XCII. **Vernon Smith**: "I think it better to say that we believe in, or have a certain faith, that there are principles, but we are still looking for them."

XCIII. **Charlie Plott**: "All three—experiments, labs, and economics—are going to change a lot. Moving small (into the mind and chemistry) and moving large (big experiments with hundreds of people and markets) can happen only if supplemented by a change in all three."

# References


1977. "Program of the Econometric Society Summer European Meeting, August 24–27, 1976, Helsinki, Finland." *Econometrica*, 45(1), 252–56.

1981. "Program of the Econometric Society World Congress." *Econometrica*, 49(1), 247–76.

1991. "Biography of Morris H. Degroot." *Statistical Science*, 6(1), 3–14.

1998. "Lawrence Fouraker, Former Business School Dean, Dies at 74," *The Harvard University Gazette*. Boston, MA: President and Fellows of Harvard College.

2002. "Preface. International Journal of Game Theory, 1995–2001." *International Journal of Game Theory*, 31, 151–53.

**Abdulkadiroglu, Atila; Parag A. Pathak and Alvin E. Roth.** 2005a. "The New York City High School Match." *The American Economic Review*, 95(2), 364–67.

**Abdulkadiroglu, Atila; Parag A. Pathak; Alvin E. Roth and Tayfun Sönmez.** 2005b. "The Boston Public School Match." *The American Economic Review. Papers and Proceedings*, 95(2), 368–71.

**Allais, M.** 1953. "Le Comportement De L'homme Rationnel Devant Le Risque: Critique Des Postulats Et Axiomes De L'ecole Americaine." *Econometrica: Journal of the Econometric Society*, 21(4), 503–46.

**Anderson, Lisa R. and Charles A. Holt.** 1996. "Information Cascades." *The Journal of Economic Perspectives*, 10(4), 187–93.

____. 1997. "Information Cascades in the Laboratory." *The American Economic Review*, 87(5), 847–62.

**Armantier, Olivier; Charles A. Holt and Charles R. Plott.** 2010. "A Reverse Auction for Toxic Assets." *Social Science Working Paper, California Institute of Technology*, 1330.

**Ayllon, Teodoro and Nathan H. Azrin.** 1968. *The Token Economy; a Motivational System for Therapy and Rehabilitation*. New York: Appleton-Century-Crofts.

**Bardsley, Nick; Robin Cubitt; Graham Loomes; Peter Moffatt; Chris Starmer and Robert Sugden.** 2010. *Experimental Economics: Rethinking the Rules*. Princeton: Princeton University Press.

**Battalio, Raymond C.; Edwin B. Fisher; John H. Kagel; Robert L. Basmann; Robin C. Winkler and Leonard Krasner.** 1974. "An Experimental Investigation of Consumer Behavior in a Controlled Environment." *Journal of Consumer Research*, 1(2).

**Battalio, Raymond C.; John H. Kagel; Robin C. Winkler; Edwin B. Fisher; Robert L. Basmann and Leonard Krasner.** 1973. "A Test of Consumer Demand Theory Using Observations of Individual Consumer Purchases." *Economic Inquiry*, 11(4), 411–28.

**Bazerman, Max H. and William F. Samuelson.** 1983. "I Won the Auction but Don't Want the Prize." *Journal of Conflict Resolution*, 27(4), 618–34.




A. Svorenčík, H. Maas (eds.), *The Making of Experimental Economics*,
DOI 10.1007/978-3-319-20952-4

**Becker, Gary; Maurice H. DeGroot and Jacob Marschak.** 1964. "Measuring Utility by a Single-Response Sequential Method." *Behavioral Science*, 9(3), 226–32.

**Becker, Otwin.** 2010. "Encounters with Reinhard Selten: An Office Mate's Report," A. Ockenfels and A. Sadrieh, *The Selten School of Behavioral Economics*. Berlin Heidelberg: Springer, 9–17.

**Becker, Otwin and Reinhard Selten.** 1970. "Experiences with the Management Game Sinto-Market," H. Sauermann, *Beiträge zur Experimentellen Wirtschaftsforschung*. Tübingen Mohr, 136–50.

**Békésy, Georg von.** 1960. *Experiments in Hearing*. New York: McGraw-Hill.

**Binger, Brian R.; R. F. Copple and Elizabeth Hoffman.** 1995a. "Contingent Valuation Methodology in the Natural Resource Damage Regulatory Process: Choice Theory and the Embedding Phenomenon." *Natural Resources Journal*, 35(3), 443–59.

____. 1995b. "The Use of Contingent Valuation Methodology in Natural Resource Damage Assessments: Legal Fact and Economic Fiction." *Northwestern University Law Review*, 89 (3), 1029–116.

**Blackmann, J.** 1978. "Experimental Studies of Choice in Economics." *NSF Program Report*, 1 (3 (June)), pp. 7–14.

**Brewer, Paul J. and Charles R. Plott.** 1996. "A Binary Conflict Ascending Price (Bicap) Mechanism for the Decentralized Allocation of the Right to Use Railroad Tracks." *International Journal of Industrial Organization*, 14(6), 857–86.

**Brunner, Christoph; Jacob K. Goeree; Charles A. Holt and John O. Ledyard.** 2010. "An Experimental Test of Flexible Combinatorial Spectrum Auction Formats." *American Economic Journal: Microeconomics*, 2(1), 39–57.

**Bueno de Mesquita, Bruce and Kenneth A. Shepsle.** 2001. *William Harrison Riker: September 22, 1920-June 26, 1993*. Washington, D.C.: National Academy Press.

**Burtraw, Dallas; Jacob Goeree; Charles A. Holt; Erica Myers; Karen Palmer and William Shobe.** 2009. "Collusion in Auctions for Emission Permits: An Experimental Analysis." *Journal of Policy Analysis and Management*, 28(4), 672–91.

____. 2010a. "The Effects of a Loose Cap on Emissions Permits: An Experimental Analysis of Alternative Auction Formats." *Agricultural and Resource Economics Review*, 39(2), 162–75.

____. 2010b. "Price Discovery in Auctions for Emissions Permits: A Comparison of Sealed Bid, Multi-Round, and Continuous Auctions," M. R. Isaac and D. A. Norton, *Experiments on Energy, the Environment, and Sustainability, Research in Experimental Economics, Vol. 14*. Bingley:UK, : Emerald Group Publishing, 11–36.

**Cartwright, Dorwin and Alvin Frederick Zander.** 1960. *Group Dynamics: Research and Theory*. Evanston, Ill: Row, Peterson.

**Castore, Carl H.** 1972. *Intragroup Concordance and the Effectiveness of Majority Rule Decisions*. Ft. Belvoir: Defense Technical Information Center. Purdue Univ Lafayette.

**Castore, Carl H. and J. Keith Murnighan.** 1973. *Decision Rule and Intragroup Goal Concordance as Determinants of Individual Reactions to and Group Decisions*. Ft. Belvoir: Defense Technical Information Center. Purdue Univ Lafayette.

____. 1978. "Determinants of Support for Group Decisions." *Organizational Behavior and Human Performance*, 22(1), 75–92.

**Chao, Hung-po and Hillard G. Huntington.** 1998. *Designing Competitive Electricity Markets*. Dordrecht: Kluwer Academic Publishers.

**Cherry, Todd L.; Peter Frykblom and Jason F. Shogren.** 2002. "Hardnose the Dictator." *The American Economic Review*, 92(04), 1218–21.

**Cohen, L. and Charles R. Plott.** 1978. "Communication and Agenda Influence: The Chocolate Pizza Design," H. Sauermann, *Coalition Forming Behavior*. Tübingen: Mohr, 329–57.

**Cooper, David J.; John H. Kagel; Wei Lo and Qing Liang Gu.** 1999. "Gaming against Managers in Incentive Systems: Experimental Results with Chinese Students and Chinese Managers." *The American Economic Review*, 89(4), 781–804.

**Coppinger, Vicki M.; Vernon L. Smith and Jon A. Titus.** 1980. "Incentives and Behavior in English, Dutch and Sealed-Bid Auctions." *Economic Inquiry*, 18(1), 1–22.

**Cummings, R.; Susan Laury and Charles A. Holt.** 2004. "Using Laboratory Experiments for Policy Making: An Example from the Georgia Irrigation Reduction Auction." *Journal of Policy Analysis and Management*, 3(2), 341–63.

**Cyert, R. M.; J. G. March and W. H. Starbuck.** 1961. "Two Experiments on Bias and Conflict in Organizational Estimation." *Management Science*, 7(3), 254–64.

**Cyert, Richard Michael; Maurice H. DeGroot and Charles A. Holt.** 1978. "Sequential Investment Decisions with Bayesian Learning." *Management Science*, 24(7), 712–18.

____. 1979. "Capital Allocation within a Firm." *Syst. Res. Behavioral Science*, 24(5), 287–95.

**Cyert, Richard Michael and James G. March.** 1963. *A Behavioral Theory of the Firm*. Englewood Cliffs, N.J.: Prentice-Hall.

**Davis, Douglas D. and Charles A. Holt.** 1993. "Experimental Economics," Princeton University Press, 406–26.

**Denton, Michael J.; Stephen J. Rassenti; Vernon L. Smith and Steven R. Backerman.** 2001. "Market Power in a Deregulated Electrical Industry." *Decision Support Systems*, 30(3), 357–81.

**Dolbear, F. Trenery.** 1963. "Individual Choice under Uncertainty: An Experimental Study." *Yale Economic Essays*, 3(2), 419–69.

**Dolbear, F. Trenery and Lester B. Lave.** 1966. "Risk Orientation as a Predictor in the Prisoner's Dilemma." *The Journal of Conflict Resolution*, 10(4), 506–15.

**Dolbear, F. Trenery; Lester B. Lave; G. Bowman; A. Lieberman; Edward Prescott; F. Rueter and Roger Sherman.** 1968. "Collusion in Oligopoly: An Experiment on the Effect of Numbers and Information." *The Quarterly Journal of Economics*, 82(2), 240–59.

**Eckel, Catherine C. and Philip J. Grossman.** 1996. "Altruism in Anonymous Dictator Games." *Games and Economic Behavior*, 16(2), 181–91.

**Edwards, Ward.** 1954. "Variance Preferences in Gambling." *The American Journal of Psychology*, 67(3), 441–52.

**Engvall Siegel, Alberta.** 1964. "Sidney Siegel: A Memoir," S. Messick and A. H. Brayfield, *Decision and Choice. Contributions of Sidney Siegel*. New York McGraw-Hill Book Co, 1–23.

**Ferejohn, John A.** 1979. "An Experimental Analysis of Decision-Making Procedures for Discrete Public Goods: A Case Study of a Problem in Institutional Design," V. L. Smith, *Research in Experimental Economics*. Greenwich, Connecticut: JAI Press, 1–58.

**Ferejohn, John A.; Robert Forsythe; Roger G. Noll and Thomas R. Palfrey.** 1982. "An Experimental Examination of Auction Mechanisms for Discrete Public Goods," *Research in Experimental Economics*. Greenwich, Connecticut: JAI Press.

**Ferejohn, John A. and Roger G. Noll.** 1976. "An Experimental Market for Public Goods: The Pbs Station Program Cooperative." *The American Economic Review*, 66(2), 267–73.

**Fiorina, Morris P. and Charles R. Plott.** 1978. "Committee Decisions under Majority Rule: An Experimental Study." *The American Political Science Review*, 72(2), 575–98.

**Forsythe, Robert; Thomas R. Palfrey and Charles R. Plott.** 1982. "Asset Valuation in an Experimental Market." *Econometrica*, 50(3), 537–67.

**Fouraker, Lawrence E. and Sidney Siegel.** 1961. Bargaining Behavior. Volume 2. Experiments in Oligopoly. Pennsylvania State University.

____. 1963. *Bargaining Behavior*. New York: McGraw-Hill.

**Friedman, Daniel and Alessandra Cassar.** 2004. *Economics Lab: An Intensive Course in Experimental Economics*. London: Taylor & Francis Ltd.

**Friedman, Daniel and Shyam Sunder.** 1994. *Experimental Methods: A Primer for Economists*. Cambridge [England]; New York: Cambridge University Press.

**Friedman, James W.** 1963. "Individual Behavior in Oligopolistic Markets: An Experimental Study." *Yale Economic Essays*, 3(2), 359–417.

____. 1967. "An Experimental Study of Cooperative Duopoly." *Econometrica*, 35(3/4), 379–97.

____. 1971. "A Non-Cooperative Equilibrium for Supergames." *The Review of Economic Studies*, 38(1), 1–12.

____. 1973. "A Non-Cooperative Equilibrium for Supergames: A Correction." *The Review of Economic Studies*, 40(3), 435.

**Friedman, James W. and Austin Curwood Hoggatt.** 1980. *An Experiment in Noncooperative Oligopoly*. Greenwich, Conn.: JAI Press.

**Fromkin, Howard L.** 1969. "The Behavioral Science Laboratories at Purdue's Krannert School." *Administrative Science Quarterly*, 14(2), 171–77.

**Goeree, Jacob; Charles A. Holt and John O. Ledyard.** 2004. "An Experimental Investigation of the Threshold Problem with Hierarchical Package Bidding," Wireless Telecommunications Bureau of the Federal Communications Commission.

____. 2006. "An Experimental Comparison of the Fcc's Combinatorial and Non-Combinatorial Simultaneous Multiple Round Auctions," Wireless Telecommunications Bureau of the Federal Communications Commission.

____. 2007. "An Experimental Comparison of Flexible and Tiered Package Bidding," Wireless Telecommunications Bureau of the Federal Communications Commission.

**Goeree, Jacob K. and Charles A. Holt.** 2010. "Hierarchical Package Bidding: A Paper & Pencil Combinatorial Auction." *Games and Economic Behavior*, 70(1), 146–69.

**Goeree, Jacob K.; Charles A. Holt; Karen Palmer; William Shobe and Dallas Burtraw.** 2010. "An Experimental Study of Auctions Versus Grandfathering to Assign Pollution Permits." *Journal of the European Economic Association*, 8(2–3), 514–25.

**Grether, David M.; R. Mark Isaac and Charles R. Plott.** 1989 [1979]. *The Allocation of Scarce Resources : Experimental Economics and the Problem of Allocating Airport Slots*. Boulder: Westview Press.

**Grether, David M. and Charles R. Plott.** 1979. "Economic Theory of Choice and the Preference Reversal Phenomenon." *The American Economic Review*, 69(4), 623–38.

____. 1984. "The Effects of Market Practices in Oligopolistic Markets: An Experimental Examination of the Ethyl Case." *Economic Inquiry*, 22(4), 479–507.

**Groves, Theodore and John O. Ledyard.** 1977. "Optimal Allocation of Public Goods: A Solution to the "Free Rider" Problem." *Econometrica*, 45(4), 783–809.

**Guala, Francesco.** 2005. *The Methodology of Experimental Economics*. Cambridge; New York: Cambridge University Press.

**Güth, Werner; Rolf Schmittberger and Bernd Schwarze.** 1982. "An Experimental Analysis of Ultimatum Bargaining." *Journal of Economic Behavior & Organization*, 3(4), 367–88.

**Harrison, Glenn W. and Kevin A. Mccabe.** 1996. "Stability and Preference Distortion in Resource Matching: An Experimental Study of the Marriage Market," M. R. Isaac, *Research in Experimental Economics. Vol. 8*. Greenwich, CT: JAI Press.

**Harstad, Ronald M.** 1996. "Special Issue on Laboratory Investigations of Expectations in Games: The Amsterdam Papers." *International Journal of Game Theory*, 25(3), n.p.

**Heady, Earl O.** 1957. "An Econometric Investigation of the Technology of Agricultural Production Functions." *Econometrica*, 25(2), 249–68.

**Heukelom, Floris.** 2010. "Measurement and Decision Making at the University of Michigan in the 1950s and 1960s." *Journal of the History of the Behavioral Sciences*, 46(2), 189–207.

**Hey, John Denis.** 1991. *Experiments in Economics*. Oxford, UK; Cambridge, USA: B. Blackwell.

**Hinich, Melvin J.; John O. Ledyard and Peter C. Ordeshook.** 1972. "Nonvoting and the Existence of Equilibrium under Majority Rule." *Journal of Economic Theory*, 4(2), 144–53.

____. 1973. "A Theory of Electoral Equilibrium: A Spatial Analysis Based on the Theory of Games." *The Journal of Politics*, 35(1).

**Hoffman, Elizabeth.** 1972. "The Sources of Mortality Changes in Italy since Unification," *History*. Philadelphia: Graduate School of Arts and Sciences, University of Pennsylvania.

**Hoffman, Elizabeth.** 1979. "Essays in Optimal Resource Allocation under Uncertainty with Capacity Constraints," Pasadena: California Institute of Technology.

**Hoffman, Elizabeth.** 1981. *The Sources of Mortality Changes in Italy since Unification.* New York: Arno Press.

**Hoffman, Elizabeth; Kevin A. McCabe and Vernon L. Smith.** 1996. "On Expectations and the Monetary Stakes in Ultimatum Games." *International Journal of Game Theory*, 25(3), 289–301.

**Hoffman, Elizabeth; Kevin McCabe; Keith Shachat and Vernon L. Smith.** 1994. "Preferences, Property Rights, and Anonymity in Bargaining Games." *Games and Economic Behavior*, 7(3), 346–80.

**Hoffman, Elizabeth and Edward W. Packel.** 1984. "Stochastic Model of Coalition Formation with Exogenous Costs: Theory and Experiments," M. Holler, *Coalitions and Collective Action.* Physica Verlag 283–94.

____. 1982. "A Stochastic Model of Committee Voting with Exogenous Costs: Theory and Experiments." *Syst. Res. Behavioral Science*, 27(1), 43–56.

**Hoffman, Elizabeth and Matthew L. Spitzer.** 1982. "The Coase Theorem: Some Experimental Tests." *Journal of Law and Economics*, 25(1), 73–98.

____. 1985. "Entitlements, Rights, and Fairness: An Experimental Examination of Subjects' Concepts of Distributive Justice." *The Journal of Legal Studies*, 14(2).

**Hoggatt, Austin C.; Joseph Esherick and John T. Wheeler.** 1969. "A Laboratory to Facilitate Computer-Controlled Behavioral Experiments." *Administrative Science Quarterly*, 14(2), 202–07.

**Hoggatt, Austin Curwood; James W. Friedman and Shlomo Gill.** 1975. *Price Signalling in Experimental Oligopoly.* [S.l.]: University of Rochester, Dept. of Economics.

**Hoggatt, Austin C.; James W. Friedman and Shlomo Gill.** 1976. "Price Signaling in Experimental Oligopoly." *The American Economic Review*, 66(2), 261–66.

**Holt, Charles A.** 2006. *Markets, Games, and Strategic Behavior: Recipes for Interactive Learning.* Boston: Pearson Addison Wesley.

**Holt, Charles A. and Roger Sherman.** 1994. "The Loser's Curse." *The American Economic Review*, 84(3), 642–52.

**Holt, Charles A.; William Shobe; Dallas Burtraw; Karen Palmer and Jacob Goeree.** 2007. "Auction Design for Selling Co2 Emissions Allowances under the Regional Greenhouse Gas Initiative."

**Hong, James T. and Charles R. Plott.** 1982. "Rate Filing Policies for Inland Water Transportation: An Experimental Approach." *The Bell Journal of Economics*, 13(1), 1–19.

**Hughes, G. David and Philippe A. Naert.** 1970. "A Computer-Controlled Experiment in Consumer Behavior." *The Journal of Business*, 43(3), 354–72.

**Isaac, Mark R. and Charles R. Plott.** 1978. "Cooperative Game Models of the Influence of the Closed Rule in Three Person, Majority-Rule Committees: Theory and Experiment," P. Ordeshook, *Game Theory and Political Science.* New York: NUY Press, NA.

**Jorgenson, Dale W. and Calvin D. Siebert.** 1968. "A Comparison of Alternative Theories of Corporate Investment Behavior." *The American Economic Review*, 58(4), 681–712.

**Kagel, John H.** 1972. "Token Economies and Experimental Economics." *Journal of Political Economy*, 80(4), 779–85.

**Kagel, John H.; Raymond C. Battalio; Howard Rachlin; Leonard Green; Robert L. Basmann and W. R. Klemm.** 1975. "Experimental Studies of Consumer Demand Behavior Using Laboratory Animals." *Economic Inquiry*, 13(1), 22–38.

**Kagel, John H.; Ronald M. Harstad and Dan Levin.** 1987. "Information Impact and Allocation Rules in Auctions with Affiliated Private Values: A Laboratory Study." *Econometrica*, 55(6), 1275–304.

**Kagel, John H. and Dan Levin.** 1985. "Individual Bidder Behavior in First-Price Private Value Auctions." *Economics Letters*, 19(2), 125–28.

____. 1986. "The Winner's Curse and Public Information in Common Value Auctions." *The American Economic Review*, 76(5), 894–920.

____. 2002. *Common Value Auctions and the Winner's Curse.* Princeton, N.J.: Princeton University Press.

**Kagel, John H. and Alvin E. Roth.** 1995. *The Handbook of Experimental Economics*. Princeton, N.J.: Princeton University Press.

____. 2000. "The Dynamics of Reorganization in Matching Markets: A Laboratory Experiment Motivated by a Natural Experiment." *Quarterly Journal of Economics*, 115(1), 201–35.

**Kahneman, Daniel and Amos Tversky.** 1979. "Prospect Theory: An Analysis of Decision under Risk." *Econometrica*, 47(2), 263–91.

**Kalisch, Gerhard K.; J. W. Millnor; John F. Nash and E. D. Nering.** 1954. "Some Experimental N-Person Games," R. M. Thrall, C. H. Coombs and R. L. Davis, *Decision Processes*. New York: Wiley, 513–18.

**Kirchler, Michael; Jürgen Huber and Thomas Stöckl.** 2012. "Thar She Bursts: Reducing Confusion Reduces Bubbles." *The American Economic Review*, 102(2), 865–83.

**Kloek, Teun.** 2001. "Obituary: Henri Theil, 1924–2000." *Statistica Neerlandica*, 55(3), 263–69.

**Krech, David; Richard S. Crutchfield and Egerton L. Ballachey.** 1962. *Individual in Society : A Textbook of Social Psychology*. Tokyo [etc.]: McGraw-Hill Kogokusha.

**Lave, Lester B.** 1960. *An Empirical Description of the Prisoner's Dilemma Game*. Santa Monica, Calif.: Rand Corporation.

____. 1962. "An Empirical Approach to the Prisoners' Dilemma Game." *The Quarterly Journal of Economics*, 76(3), 424–36.

____. 1965. "Factors Affecting Co-Operation in the Prisoner's Dilemma." *Syst. Res. Behavioral Science*, 10(1), 26–38.

**Leffler, George L.** 1951. *The Stock Market*. New York: Ronald Press.

**Lei, Vivian; Charles N. Noussair and Charles R. Plott.** 2001. "Nonspeculative Bubbles in Experimental Asset Markets: Lack of Common Knowledge of Rationality Vs. Actual Irrationality." *Econometrica*, 69(4), 831–59.

**Leonard, Robert J.** 2010. *Von Neumann, Morgenstern, and the Creation of Game Theory: From Chess to Social Science, 1900–1960*. New York: Cambridge University Press.

**Levitt, Steven D. and John A. List.** 2007. "What Do Laboratory Experiments Measuring Social Preferences Reveal About the Real World." *The Journal of Economic Perspectives*, 21(02), 153–74.

**Lim, Suk S.; Edward C. Prescott and Shyam Sunder.** 1994. "Stationary Solution to the Overlapping Generations Model of Fiat Money: Experimental Evidence." *Empirical Economics*, 19(2), 255–77.

**Lloyd, Cliff.** 1967. *Microeconomic Analysis*. Homewood, Ill.: R.D. Irwin.

____. 1980. *The Collected Works of Cliff L. Lloyd*. Burnaby, B.C.: School of Business Administration and Economics, Simon Fraser University.

**Lucas, Robert E., Jr.** 1986. "Adaptive Behavior and Economic Theory." *The Journal of Business*, 59(4), 401–26.

**Maschler, Michael.** 1965. *Playing an N-Person Game, an Experiment*. Princeton, N.J.: Princeton University, Econometric Research Program Research Memorandum No. 73.

____. 1972. "Equal Share Analysis of Characteristic Function Experiments," H. Sauermann, *Contributions to Experimental Economics = Beiträge zur Experimentellen Wirtschaftsforschung*. Tübingen: Mohr, 130–65.

____. 1978. "Playing an N-Person Game: An Experiment," H. Sauermann, *Bargaining Behavior. Contributions to Experimental Economics = Beiträge zur Experimentellen Wirtschaftsforschung*. Tübingen: Mohr, 231–328.

**May, Kenneth O.** 1954. "Intransitivity, Utility, and the Aggregation of Preference Patterns." *Econometrica*, 22(1), 1–13.

**Miller, Ross M.; Charles R. Plott and Vernon L. Smith.** 1977. "Intertemporal Competitive Equilibrium: An Empirical Study of Speculation." *The Quarterly Journal of Economics*, 91(4), 599–624.

**Morgenstern, Oskar.** 1963. *On the Accuracy of Economic Observations*. Princeton, N.J.: Princeton University Press.

**Mosteller, Frederick; Stephen E. Fienberg and David C. Hoaglin.** 2006. *Selected Papers of Frederick Mosteller*. New York: Springer.

**Mosteller, Frederick; Stephen E. Fienberg; David C. Hoaglin and Judith M. Tanur.** 2010. *The Pleasures of Statistics : An Autobiography of Frederick Mosteller*. New York: Springer.

**Mosteller, Frederick and Philip Nogee.** 1951. "An Experimental Measurement of Utility." *The Journal of Political Economy*, 59(5), 371–404.

**Mount, Kenneth and Stanley Reiter.** 1974. "The Informational Size of Message Spaces." *Journal of Economic Theory*, 8(2), 161–92.

**Murnighan, J. Keith and Alvin E. Roth.** 1980. "Effects of Group Size and Communication Availability on Coalition Bargaining in a Veto Game." *Journal of Personality and Social Psychology*, 39(1), 92–103.

**Ockenfels, Axel and Abdolkarim Sadrieh.** 2010. "The Selten School of Behavioral Economics a Collection of Essays in Honor of Reinhard Selten," Berlin; Heidelberg: Springer.

**Olson, Mark; Stephen Rassenti; Mary Rigdon and Vernon Smith.** 2003. "Market Design and Human Trading Behavior in Electricity Markets." *IIE Transactions*, 35(9), 833–49.

**Owen, Guillermo.** 2010. "Michael Maschler's Bibliography." *International Journal of Game Theory. Special Issue in Honor of Michael Maschler*, 39(1–2), 301–08.

**Pecorino, Paul and Mark Van Boening.** 2010. "Fairness in an Embedded Ultimatum Game." *Journal of Law and Economics*, 53(2), 263–87.

**Plott, Charles R.** 1976. "Axiomatic Social Choice Theory: An Overview and Interpretation." *American Journal of Political Science*, 20(3), 511–96.

**____.** 1979. "The Application of Laboratory Experimental Methods to Public Choice," C. S. Russell, *Collective Decision-Making: Applications from Public Choice Theory*. Baltimore, Md.: Johns Hopkins Press for Resources for the Future, 137–60.

**____.** "Plott Papers," *personal archive of Charles Plott*. California Institute of Technology,

**____.** 1981. "Theories of Industrial Organization as Explanations of Experimental Market Behavior," S. C. Salop, *Strategy, Predation, and Antitrust Analysis*. Federal Trade Commission.

**____.** 1982. "Industrial Organization Theory and Experimental Economics." *Journal of Economic Literature*, 20(4), 1485–527.

**____.** 1988. "Research on Pricing in a Gas Transportation Network," *Office of Economic Policy Technical Report no. 88–2*. Washington, D.C.: Federal Energy Regulatory Commission.

**____.** 2001. *Collected Papers on the Experimental Foundations of Economics and Political Science*. Cheltenham, UK; Northampton, MA, USA: Edward Elgar.

**Plott, Charles R. and Michael E. Levine.** 1978. "A Model of Agenda Influence on Committee Decisions." *The American Economic Review*, 68(1), 146–60.

**Plott, Charles R. and Vernon L. Smith.** 1978. "An Experimental Examination of Two Exchange Institutions." *The Review of Economic Studies*, 45(1), 133–53.

**Plott, Charles R. and L. L. Wilde.** 1982. "Professional Diagnosis Vs. Self-Diagnosis: An Experimental Examination of Some Special Features of Markets with Uncertainty," V. L. Smith, *Research in Experimental Economics*. Greenwich, Conn.: JAI Press.

**Princeton University, Conference.** 1961. "Recent Advances in Game Theory; Papers Delivered at a Meeting of the Princeton University Conference, October [Sic] 4–6, 1961," *Recent advances in game theory*. Princeton, N.J.

**Radford, R. A.** 1945. "The Economic Organization of P.O.W. Camps." *Economica*, 12, 189–201.

**Raghavan, T. E. S.; T. S. Ferguson; T. Parthasarathy and O. J. Vrieze** eds. 1990. *Stochastic Games and Related Topics : In Honor of Professor L.S. Shapley*. Dordrecht; Boston: Kluwer Academic Publishers.

**Rapoport, Amnon.** 1990. *Experimental Studies of Interactive Decisions*. Dordrecht; Boston: Kluwer Academic.

**Rassenti, Stephen.** 1982. "Zero/One Decision Problems with Multiple Resource Constraints Algorithms and Applications," *Systems and Industrial Engineering*. University of Arizona,

**Rassenti, S. J. and V. L. Smith.** 1998. "Deregulating Electric Power: Market Design Issues and Experiments." *International Series in Operations Research and Management Science*, (13), 105–20.

**Rassenti, Stephen J.; Vernon L. Smith and R. L. Bulfin.** 1982. "A Combinatorial Auction Mechanism for Airport Time Slot Allocation." *The Bell Journal of Economics*, 13(2), 402–17.

**Rassenti, Stephen J.; Vernon L. Smith and Bart J. Wilson.** 2002. "Using Experiments to Inform the Privatization/Deregulation Movement in Electricity." *CATO Journal*, 21, 515–44.

____. 2003. "Controlling Market Power and Price Spikes in Electricity Networks: Demand-Side Bidding." *Proceedings of the National Academy of Sciences of the United States of America*, 100(5), 2998–3003.

**Reed, Harvey Jay.** 1973. "An Experimental Study of Equilibrium in a Competitive Market," Purdue University,

**Reiley, David H.; Michael B. Urbancic and Mark Walker.** 2008. "Stripped-Down Poker: A Classroom Game with Signaling and Bluffing." *The Journal of Economic Education*, 39(4), 323–41.

**Repcheck, Jack.** 2003. *The Man Who Found Time : James Hutton and the Discovery of the Earth's Antiquity*. Cambridge, MA: Perseus.

____. 2007. *Copernicus' Secret : How the Scientific Revolution Began*. New York: Simon & Schuster.

**Ricciardi, Franc M.** 1957. *Top Management Decision Simulation: The A.M.A. Approach*. New York: American management association.

**Riedl, Arno and Frans van Winden.** 2001. "Does the Wage Tax System Cause Budget Deficits? A Macro-Economic Experiment." *Public Choice*, 109(3), 371–94.

____. 2007. "An Experimental Investigation of Wage Taxation and Unemployment in Closed and Open Economies." *European Economic Review*, 51(4), 871–900.

____. 2012. "Input Versus Output Taxation in an Experimental International Economy." *European Economic Review*, 56(2), 216–32.

**Riker, William H.** 1962. *The Theory of Political Coalitions*. New Haven: Yale University Press.

____. 1967. "Bargaining in a Three-Person Game." *The American Political Science Review*, 61(3), 642–56.

____. 1971. "An Experimental Examination of Formal and Informal Rules of a Three-Person Game," B. Lieberman, *Social Choice*. New York: Gordon and Breach,

____. 1982. *Liberalism against Populism: A Confrontation between the Theory of Democracy and the Theory of Social Choice*. San Francisco: W.H. Freeman.

**Riker, William H. and Richard G. Niemi.** 1964. "Anonymity and Rationality in the Essential Three-Person Game." *Human Relations*, 17(2), 131–41.

**Riker, William H. and Peter C. Ordeshook.** 1973. *An Introduction to Positive Political Theory*. Englewood Cliffs, N.J.: Prentice-Hall.

**Riker, William H. and William James Zavoina.** 1970. "Rational Behavior in Politics: Evidence from a Three Person Game." *The American Political Science Review*, 64(1), 48–60.

**Roth, Alvin E.** 1979. *Axiomatic Models of Bargaining*. Berlin; New York: Springer-Verlag.

____. 1986. "Laboratory Experimentation in Economics." *Economics and Philosophy*, 2, 245–73

____. 1987a. "Laboratory Experimentation in Economics," T. Bewley, *Advances in Economic Theory 1985 (Symposia of the 5th World Congress of the Econometric Society*. Cambridge University Press, 269–99.

____ ed. 1987c. *Laboratory Experimentation in Economics: Six Points of View*. Cambridge; New York: Cambridge University Press.

____ ed. 1988. *The Shapley Value : Essays in Honor of Lloyd S. Shapley*. Cambridge; New York: Cambridge University Press.

____. 1995. "Bargaining Experiments," J. H. Kagel a. A. E. Roth, *Handbook of Experimental Economics*. NJ: Princeton Univ. Press.

**Roth, Alvin E. and Michael W. Malouf.** 1979. "Game-Theoretic Models and the Role of Information in Bargaining." *Psychological Review*, 86(6), 574–94.

**Roth, Alvin E. and J. Keith Murnighan.** 1978. "Equilibrium Behavior and Repeated Play of the Prisoner's Dilemma Games." *Journal of Mathematical Psychology*, 17(2), 189–98.

____. 1982. "The Role of Information in Bargaining: An Experimental Study." *Econometrica*, 50(5).

____. 2004. "Some of the Ancient History of Experimental Economics and Social Psychology: Reminiscences and Analysis of a Fruitful Collaboration," D. De Cremer, *Social Psychology and Economics*. Mahwah, N.J: Lawrence Erlbaum, 321–34.

**Roth, Alvin E. and Elliott Peranson.** 1999. "The Redesign of the Matching Market for American Physicians: Some Engineering Aspects of Economic Design." *The American Economic Review*, 89(4), 748–80.

**Roth, Alvin E.; Vesna Prasnikar; Masahiro Okuno-Fujiwara and Shmuel Zamir.** 1991. "Bargaining and Market Behavior in Jerusalem, Ljubljana, Pittsburgh, and Tokyo: An Experimental Study." *The American Economic Review*, 81(5), 1068–95.

**Rousseas, Stephen W. and Albert G. Hart.** 1951. "Experimental Verification of a Composite Indifference Map." *The Journal of Political Economy*, 59(4), 288–318.

**Sadrieh, Abdolkarim; Klaus Abbink and Reinhard Tietz.** 2008. *Experimental Economics in Germany, Austria, and Switzerland: A Collection of Papers in Honor of Reinhard Tietz.* Marburg: Metropolis-Verl.

**Salmon, Timothy and Bart Wilson.** 2008. "Second Chance Offers Versus Sequential Auctions: Theory and Behavior." *Economic Theory*, 34(1), 47–67.

**Sauermann, Heinz** ed. 1967. *Beiträge zur Experimentellen Wirtschaftsforschung*. Tübingen: Mohr.

____. ed. 1972. *Contributions to Experimental Economics. Vol. 3 Beiträge zur Experimentellen Wirtschaftsforschung. Band. 3*. Tübingen: Mohr.

____. ed. 1978a. *Bargaining Behavior*. Tübingen: Mohr.

____. ed. 1978b. *Coalition Forming Behavior*. Tübingen: Mohr.

**Sauermann, Heinz and Reinhard Selten.** 1960. "An Experiment in Oligopoly," *General Systems, Yearbook of the Society for General Systems Research*. Ann Arbor, MI: Society for General Systems, 85–114, 206.

**Schumann, Jochen.** 1983. *Heinz Sauermann: Wirtschaftstheoretiker und Sozialwissenschaftler: Beitr*. Frankfurt am Main: Barudio [und] Hess.

**Selten, Reinhard.** 1961. "Bewertung Von N-Personenspielen," Frankfurt: Johann Wolfgang Goethe-Universität Frankfurt am Main.

____. 1965. "Spieltheoretische Behandlung eines Oligopolmodells mit Nachfrageträgheit. Teil 1 & 2 Eigenschaften des dynamischen Preisgleichgewichts" *Zeitschrift für die gesamte Staatswissenschaft*, 121, 301–24, 667–89.

____. 1991. *Game Equilibrium Models*. Berlin; New York: Springer Verlag.

**Selten, Reinhard and Heinz Sauermann.** 1959. "Ein Oligopolexperiment." *Zeitschrift für die gesamte Staatswissenschaft*, 115, 427–71.

**Siegel, Sidney.** 1953. "Certain Determinants and Correlates of Authoritarianism," *Dept. of Psychology*. Stanford University.

____. 1956. *Nonparametric Statistics for the Behavioral Sciences*. New York: McGraw-Hill.

**Siegel, Sidney and N. John Castellan.** 1988. *Nonparametric Statistics for the Behavioral Sciences*. New York: McGraw-Hill.

**Siegel, Sidney and Lawrence E. Fouraker.** 1960. *Bargaining and Group Decision Making; Experiments in Bilateral Monopoly*. New York: McGraw-Hill.

**Siegel, Sidney; Lawrence E. Fouraker and Donald I. Harnett**. 1961. Bargaining Behavior. Volume 1. The Uses of Information and Threat by Bilateral Monopolists of Unequal Strength. Pennsylvania State University.

**Simon, Herbert A.** 1955. "A Behavioral Model of Rational Choice." *The Quarterly Journal of Economics*, 69(1), 99–118.

____. 1956. "Rational Choice and the Structure of the Environment." *Psychological review*, 63(2), 129–38.

____. 1957. *Models of Man: Social and Rational; Mathematical Essays on Rational Human Behavior in Society Setting*. New York: Wiley.

**Slonim, Robert and Alvin E. Roth.** 1998. "Learning in High Stakes Ultimatum Games: An Experiment in the Slovak Republic." *Econometrica*, 66(03), 569–96.

**Smith, Vernon L.** "Smith Papers," *Vernon Lomax Smith Papers*. David M. Rubenstein Rare Book and Manuscript Library, Duke University, Durham, North Carolina.

**Smith, Vernon L; Arlington W. Williams; W. Kenneth Bratton and Michael G. Vannoni.** 1982. "Competitive Market Institutions: Double Auctions Vs. Sealed Bid-Offer Auctions." *The American Economic Review*, 72(1), 58–77.

**Smith, Vernon and Bart J. Wilson.** 2015 to appear. "Fair and Impartial Spectators in Experimental Economic Behavior," *Oxford Philosophical Concepts: Sympathy* (Ed. Eric Schliesser). Oxford: Oxford university Press.

**Smith, Vernon L.** 1962a. "Errata: An Experimental Study of Competitive Market Behavior." *The Journal of Political Economy*, 70(3), 322–23.

____. 1962c. "An Experimental Study of Competitive Market Behavior." *The Journal of Political Economy*, 70(2), 111–37.

____. 1973. "Notes on Some Literature in Experimental Economics." *Social Science Working Paper, California Institute of Technology*, 21, 1–27.

____. 1975. "Experimental Economics: Theory and Results." *Social Science Working Paper, California Institute of Technology*, 73(January), 1–16.

____. 1976. "Experimental Economics: Induced Value Theory." *The American Economic Review. Papers and Proceedings of the Eighty-eighth Annual Meeting of the American Economic Association*, 66(2), 274–79.

____. 1982. "Microeconomic Systems as an Experimental Science." *The American Economic Review*, 72(5), 923–55.

____. 2008a. *Discovery - a Memoir*. Bloomington, In: AuthorHouse.

____. 2008c. *Rationality in Economics: Constructivist and Ecological Forms*. Cambridge: Cambridge University Press.

____. 2012. "Adam Smith on Humanomic Behavior. Edited Transcript of Keynote Address, Academy of Behavioral Finance and Economics, UCLA." *Journal of Behavioral Finance & Economics*, 2(1), 1–20.

**Smith, Vernon L.; Gerry L. Suchanek and Arlington W. Williams.** 1988. "Bubbles, Crashes, and Endogenous Expectations in Experimental Spot Asset Markets." *Econometrica*, 56(5).

**Sondak, Harris and Max H. Bazerman.** 1989. "Matching and Negotiation Processes in Quasi-Markets." *Organizational Behavior and Human Decision Processes*, 44(2), 261–80.

____. 1991. "Power Balance and the Rationality of Outcomes in Matching Markets." *Organizational Behavior and Human Decision Processes*, 50(1), 1–23.

**Starbuck, William H.** 1993. "Watch Where You Step!" Or Indiana Starbuck Amid the Perils of Academe (Rated Pg)," *Management Laureates: A Collection of Autobiographical Essays*. Greenwich, Connecticut; London, England: JAI Press, 63–110.

**Starbuck, William H. and Frank M. Bass.** 1967. "An Experimental Study of Risk-Taking and the Value of Information in a New Product Context." *The Journal of Business*, 40(2), 155–65.

**Svorenčík, Andrej.** 2015. *The Experimental Turn: A History of Experimental Economics*. Ph.D. dissertation. University of Utrecht.

**Svorenčík, Andrej.** 2016. "The Sidney Siegel Tradition in Experimental Economics: The Divergence of Behavioral and Experimental Economics at the end of the 1980s," De Marchi and Marina Bianchi Economizing Mind, 1870–2015. When Economics and Psychology Met … or Didn't. Durham: Duke University Press (in Press)

**Thaler, Richard H.** 1992. *The Winner's Curse : Paradoxes and Anomalies of Economic Life*. New York; Toronto; New York: Free Press ; Maxwell Macmillan Canada; Maxwell Macmillan International.

____. 2015. *Misbehaving: The Making of Behavioral Economics*. W. W. Norton & Company.

**Thurstone, Leon L.** 1931. "The Indifference Function." *Journal of Social Psychology*, 2, 139–67.

**Tietz, Reinhard.** 1973. *Ein anspruchsanpassungsorientiertes Wachstums- und Konjunkturmodell (Kresko)*. Tübingen: Mohr.

**____.** 1983. *Aspiration Levels in Bargaining and Economic Decision Making: Proceedings of the Third Conference on Experimental Economics, Winzenhohl, Germany, August 29-September 3, 1982*. Berlin; Heidelberg; New York; Tokyo Springer-Verlag.

**van Winden, Frans.** 1983. *On the Interaction between State and Private Sector: A Study in Political Economics*. Amsterdam; New York: North-Holland Pub. Co.

**van Winden, Frans; Arno Riedl; J. Wit and F. van Dijk.** 2001. "An Experimental Study of the Van Elswijk Plan: Value Added Taxation Instead of Wage Taxation as a Means to Finance Unemployment Benefits," Amsterdam: CREED, University of Amsterdam.

**Wallis, W. Allen and Milton Friedman.** 1942. "The Empirical Derivation of Indifference Functions," O. Lange, F. McIntyre and T. Yntema, *Studies in Mathematical Economics and Econometrics in Memory of Henry Schultz*. Chicago: University of Chicago Press, 175–89.

**Wilde, Louis L.** 1981. "On the Use of Laboratory Experiments in Economics," J. C. Pitt, *Philosophy in Economics*. Amsterdam: Reidel, 137–48.

**Wilson, Robert.** 1977. "A Bidding Model of Perfect Competition." *The Review of Economic Studies*, 44(3), 511–18.

**Williams, Arlington W.** 2008. "Chapter 29 Price Bubbles in Large Financial Asset Markets," R. P. Charles and L. S. Vernon, *Handbook of Experimental Economics Results*. Elsevier, 242–46.

**____.** 1985. "Incentive Efficiency of Double Auctions." *Econometrica*, 53(5).

**Yaari, Menahem E.** 1965. "Convexity in the Theory of Choice under Risk." *The Quarterly Journal of Economics*, 79(2), 278–90.

The manufacturer's authorised representative in the EU is Springer Nature Customer Service Centre GmbH, Europaplatz 3, 69115 Heidelberg, Germany. If you have any concerns regarding our products, please contact ProductSafety@springernature.com

Printed and bound by CPI Group (UK) Ltd, Croydon, CR0 4YY

30/09/2026

02225651-0006